PASS YOUR AMATEUR RADIO EXTRA CLASS TEST – THE EASY WAY
2024-2028 Edition

I0095359

By: Craig E. "Buck," K4IA

ABOUT THE AUTHOR: "Buck," as we know him on the air, was first licensed in the mid-sixties as a young teenager. Today, he holds an Amateur Extra Class Radio License. Buck is an active instructor and a Volunteer Examiner. The Rappahannock Valley Amateur Radio Club named him Elmer (Trainer) of the Year three times. Buck is on the DXCC Honor Roll and has an 8 Band DXCC endorsement.

Email: K4IA@EasyWayHamBooks.com

Published by EasyWayHamBooks
130 Caroline St. Fredericksburg, Virginia 22401

EasyWayHamBooks by Craig Buck are available at Ham Radio Outlet, GigaParts, Barnes & Noble and Amazon
Pass Your Amateur Radio Technician Class Test
Pass Your Amateur Radio General Class Test
How to Chase, Work & Confirm DX
How to Get on HF
Pass Your GROL Test
Prepper Communications

Ver 2.5 Includes Errata released Jan 31, 2024

ISBN 979-8-9856739-3-7

Library of Congress Control Number: 2024900697

PASS YOUR AMATEUR RADIO EXTRA CLASS TEST – THE EASY WAY

TABLE OF CONTENTS

INTRODUCTION

There are many books and methods to study for the amateur radio exams. Most run through all the questions and four possible answers on the multiple-choice test. There are three wrong answers for each right answer; 1,809 wrong answers and 603 correct answers. No wonder people get overwhelmed. This book is unique. There are no confusing wrong answers.

The questions are roughly in the order they appear in the question pool. I have rearranged a few to keep topics together. Please excuse any tortured grammar or bad punctuation in the bold print questions and answers. They are word-for-word from the pool.

You will pass the Extra Class test, but passing won't teach much about how to operate, choose equipment and put up antennas. Be sure to check out my other books: "***How to Get on HF – The Easy Way***" for detailed instructions to get on the air and "***How to Chase, Work & Confirm DX – The Easy Way.***"

THE TEST

This book is for tests after June 30, 2024 and before July 1, 2028. The test is 50 questions from a pool of 603. The pool is large, but only about one of a dozen pool questions will be on your exam - one from each of the 50 test-subject groups. You won't get more than one question from each group.

The Extra test is harder than the General, but you can do it. Don't be intimidated. This book teaches to recognize the correct answer, even if you can't recite it. Go through this material learning to identify the correct answer, not memorize it word-for-word.

There is no Morse Code required for any class of amateur license. Morse Code is still very much alive and well on the amateur bands, and there are many reasons to learn it in the future.[1]

The good news is that we know the exact pool of questions and answers. You know what to study. If "200" is the answer in the question pool, "200" will also be the answer on your exam. Answer "A" may not be the same as answer "A" on your test, so don't memorize the answer's letter.

There is only one question from each of the 50 groups. Do your best, but if there is a question or concept you can't understand, relax, chances are good it won't be on your test. Ninety-two percent of the questions in the pool will not. Visit my Facebook group "Ham Radio Exams" to ask for help.

---

[1] See the bonus material at the end of this book and check out my book "How to Chase, Work and Confirm DX – The Easy Way." I discuss the advantages of CW over voice modes in detail.

The minimum passing score is 74%. Answer thirty-seven of the fifty questions to pass. That is only 6% of the total pool. Feel better?

Most importantly, when the day comes, take the test, even if you are uncertain. You will be better prepared than you think.

Do you know what they call the person who graduated last in his class from medical school? "Doctor" – the same as the one who graduated first in the class. My point is: pass, and no one will know the difference.

The questions are multiple-choice, and the answers are on the answer sheet. You only have to <u>recognize</u> them. That is a tremendous advantage for the test-taker. However, studying the multiple-choice format is challenging because you get bogged down and confused by seeing three wrong answers for every correct one.

The best way to study for a multiple-choice exam is to concentrate on the correct answers. That is the focus of this book. The correct response will jump out when you take the test. The wrong answers will seem strange and unfamiliar, as if in a foreign language.

Don't over-think or over-analyze. You should recognize the right answer without thinking. If you don't remember the answer, try to eliminate the obviously wrong answers and guess. There is no penalty for guessing wrong.

Here's the ultimate secret cheat: many times, the answer is a matter of common sense and logic. Often the question gives away the answer. Many of the hints and cheats in this book exploit this. Try to find some of your own memory devices.

THE TEST

The test-day routine is the same as the General test. Bring a picture ID and your FRN Number. The FRN is used only for your communications with the FCC. Search "FRN FCC" to get the link to the FCC website where you can register. If you are already licensed, the FCC has assigned you an FRN. Look it up on FCC.Gov and search in the ham database for your last name, first name, or call sign.

Bring a pen and pencil and the exam fee (usually around $15). Check ahead to get the exact amount and if the VE team prefers cash or a check.

Clear any calculator memories. The examiners will check. Turn off your cell phone, and don't look at it during the exam. You won't be able to use the cell phone calculator because the examiners can't be sure you aren't looking for answers on the Internet.

Fill out the FCC Form 605 application. Get one from the FCC website to save time on test day. It will require an email address that becomes your official contact medium for the FCC. Keep it current, as the FCC presumes anything emailed to that address is received. Failure to respond could cause license cancellation.

The Volunteer Examiner team will give you a test booklet. Ask them for the easy one – that's good for a cheap laugh. They do not know which questions are in which booklet. You also get a multiple-choice, fill-in-the-circle answer sheet. Take notes and calculate on the back of the answer sheet – not in the booklet. The VEs re-use the booklets.

Hint: Do a "memory dump" on the back of the answer sheet before you get started. Refer to it later.

The VE team will grade the exam while you wait and give a pass/fail result. Don't ask them to review the questions or tell what questions you missed. They don't know and don't have time to look. There is a bit more paperwork when you pass and walk out with a CSCE – a Certificate of Successful Completion of Examination. The CSCE for passing Extra, allows you to operate as an Extra even though not yet in the FCC database.

Add the special identifier "AE" after your call sign to use Extra Class privileges while waiting for the FCC to post the upgrade. That would be W3ABC/AE. W3ABC stroke/slash/slant AE. *Hint: Awaiting Extra.* After you appear in the database, drop the AE.

The FCC does not mail paper licenses or notices. Print a "Reference Copy" from your ULS page on the FCC website.

You may not need to take the Extra Class test if you are beyond the two-year grace period for a test-free renewal but can show you once held an Extra class license, which the FCC did not revoke. You must only pass the current Technician Class test. Check with the local VE team, in advance, for details.

HOW YOU SHOULD STUDY

Ace a multiple-choice exam by recognizing the correct answers. The EasyWayHamBooks method never shows the wrong answers, so the correct answers should pop out on test day.

There are three paths to the method. First, a narrative answering every question on the amateur radio Extra Class exam. **The test questions and answers are in bold print.** *The key words for an answer are in italic bold for focus.* Hints and cheats to decipher the questions and answers are in italic print.

Second, each chapter ends with a condensed one or two-line summary of every question and answer.

Third, is a Quick Review; no narrative, only the full question and the right answer.

Don't be discouraged if you cannot recall the answers word-for-word. The goal is to recognize them when seen on the multiple-choice test. The Summaries and Quick Review will be much more productive than practice tests.

Many of the questions and answers are long and involved, full of unneeded information. Ditch the bloat and find a simple hook to identify the answer. No need to memorize the complete answer, just enough of it to recognize. Sometimes, all you need to do is tie two words together. Look for those. Add your own hints and cheats to the ones in this book.

I do not recommend practice exams. Seeing those wrong answers, some of which are nonsense, is confusing and does not help you learn or pass. Take the practice test to build confidence, if you must, but study this book. Recognize the correct answers first.

E1 – COMMISSION'S RULES

E1A Frequency Privileges[2]

A signal is wider than you think. Be careful not to get too close to the band edges. That is the lesson of the first series of questions.

It is not legal to transmit a 3 kHz USB signal with a carrier frequency of 14.348 MHz because *the upper 1 kHz of the signal is outside the 20-meter band.*
Upper sideband puts the signal above the carrier frequency. Here, the upper limit is 3 kHz above. 14.348 MHz +3 kHz is 14.351 MHz. 14.350 is the top of the band.

The frequency that represents the lowest LSB emission totally within the band would be *3 kHz above the lower band edge.*
Lower sideband is below the carrier, so stay 3 kHz above the lower band edge.

The highest legal carrier frequency on the 20-meter band for transmitting a 2.8 kHz wide USB data signal is *14.1472 MHz.* *Hint: Too much information! Start with the highest answer and work down, adding 2.8 kHz each time. The first result below 14.350 MHz is correct.*

If an Extra Class operator hears a station calling CQ on 3.601 MHz LSB, it is NOT legal to return the call because *your sideband would extend beyond the phone band edge.* The band edge for

---

[2] (E1A) refers to the question pool subelement and group. You get one question from each group. 50 groups, 50 questions.

voice is 3.6 MHz. Your LSB signal would extend below that.

Physical control of an amateur station on a vessel or craft registered in the United States must be with any person holding an FCC-issued amateur radio license or who is *authorized for alien reciprocal operation.* Hint: Look for "alien reciprocal operation."

On 60 meters, set the CW carrier frequency at the *center frequency* of the channel. Sixty-meters is the only band restricting amateurs to channels.

The maximum power permitted on the 2200-meter band is *1 watt* EIRP (Equivalent isotropic radiated power).

Except in some parts of Alaska, the maximum power permitted on the 630-meter band is *5 watts* EIRP.

Message forwarding stations (like Packet) are usually automated. **If a station forwards a message that violates FCC rules, *the control operator of the originating station* is accountable for the rules violation.**

You can operate from a ship or airplane, but you *must obtain permission from the master* of the ship or pilot in command of the aircraft. You need permission from the boss.

On a US registered vessel in international waters, *any FCC-issued amateur license will do*. No special class or endorsements required.

E1B Station Restrictions

A spurious emission is an emission *outside its necessary bandwidth* that can be reduced or eliminated without affecting the information transmitted. *Hint: Spurious is outside its necessary bandwidth.*

An acceptable bandwidth for digital voice or slow scan TV is *3 kHz.* It should be about the same width as SSB.

You must protect an FCC monitoring facility from harmful interference if you are *within 1 mile.*

If a radiolocation system experiences interference from a repeater, the control operator *must cease operation or make changes to mitigate the interference.* Interfering with radiolocation systems can cause accidents.

The National Radio Quiet Zone is an area surrounding the *National Radio Astronomy Observatory.* *Cheat: "National is in both the question and answer.*

Before installing an antenna at a site near a public use airport, you may have to *notify the Federal Aviation Administration* and register it with the FCC. *Hint: Notify and register.*

PRB-1 applies to *state and local zoning.* State and local governments must *"reasonably accommodate"* amateur radio. No such restrictions apply to homeowner associations. They can be unreasonable.

If an amateur station's signal causes interference to domestic broadcast reception, *the FCC may place limitations* to avoid

transmitting during certain hours on frequencies that cause the interference. The FCC can impose "quiet times."

RACES is the Radio Amateur Civil Emergency Service and is part of a protocol established by FEMA and the FCC. **Any FCC-licensed amateur station *certified by the responsible civil defense organization* for the area served may be operated under RACES rules.** *Hint: RACES certifies.*

The frequencies authorized for amateur stations under RACES rules are *all amateur service frequencies* authorized to the control operator. Nothing special.

E1C Control

The maximum bandwidth for data emissions on 60 meters is *2.8 kHz*. About the same as SSB voice.

The types of communication that may be transmitted to amateur stations in foreign countries are communications incidental to the purpose of the amateur service and *remarks of a personal nature*. Same as domestic.

Power-line communication (PLC) carries data on power lines used for electric power distribution to consumers. Operation on 630 or 2200 meters could interfere. **When operating in the 2200 or 630-meter bands, you may operate *30 days* after notifying the Utilities Technology Council (UTC) before you can transmit, provided you have not been told your station is within 1 km of a PLC system using those frequencies.** *Hint: Too complicated. Look for the answer with "wait 30 days."*

Before operating, you must *notify UTC* of your call sign and the coordinates of your station.

An IARP is an International Amateur Radio Permit that allows US amateurs to operate in *certain countries of the Americas.* It is a multi-country license.

CEPT is an agreement for US amateurs to operate in *European countries* and vice versa.

To operate within the CEPT rules, you must *bring a copy of FCC Public Notice* DA 16-1048. It lists the countries and rules in several languages. *Cheat: CEPT requires a copy of "FCC Public Notice."*

A station may transmit third-party communication under automatic control *only when transmitting RTTY or data* emissions.

The maximum allowed duration of a remotely controlled station's transmissions if its control link malfunctions is *3 minutes.* It must time-out after 3 minutes.

The highest modulation index permitted for angle modulation below 29 MHz is *1.* Angle modulation is FM. The FM Modulation index is equal to the ratio of the frequency deviation to the modulating frequency. A ratio of 1 is narrow.

The permitted mean power of any spurious emission is *-43 dB* below the fundamental emission. *Hint: The question is too complicated. Remember, spurious emissions must be 43 dB below.*

Phone emissions are permitted in the *entire 630-meter band.*

E1D Amateur Space and Earth Stations

Telemetry is *one-way transmission of measurements*. *Hint: Telemetry is metering.*

***Telecommand signals* from a space telecommand station may transmit encrypted messages.** The beeps and pulses don't mean anything to a listener.

A space telecommand station is a station that transmits communications to *initiate, modify, or terminate functions* of a space station. *Hint: It commands the space station to do something.*

A balloon-borne telemetry station must identify with a *call sign*. *Hint: "Must" identify. Other information is not required of anyone.*

A station being operated by telecommand on or within 50 km of the Earth's surface must post at the station location:
- **A photocopy of the station license**
- **A label with the name, address and telephone number of the station licensee**
- **A label with the name address and telephone number of the control operator.**
- **All these choices are correct.**

All this information would help find the owner.

The maximum power permitted when operating a model craft by telecommand is *1 watt*.

The HF bands authorized for space stations are *40 m, 20 m, 15 m, and 10 m bands*. *Hint: All of 40-10 except WARC bands.*

On VHF, *only 2 meters* is available for space stations.

On UHF, *70 cm and 13 cm* are available for space stations.

Any amateur station *designated by the space station licensee* is eligible to be a telecommand station. *Hint: You need permission from the boss, the space station licensee.*

Stations eligible to operate as Earth stations are *any amateur station*, subject to the privileges of the control operator.

The stations that may transmit one-way communications are a space station, beacon station or *telecommand station*. *Hint: Look for the answer with telecommand, one-way control communications.*

E1E Volunteer Examiner Program

Count yourself fortunate that you do not have to travel to an FCC Field Office to take your test. The FCC turned testing responsibilities over to Volunteer Examiner Coordinators (VECs) in 1984. There are about 14 Volunteer Examiner Coordinator organizations, and they jointly write the question pools and administer the system. On test day, the folks you see are Volunteer Examiners (VEs), accredited by the Coordinators (VECs).

VEs and VECs may be reimbursed for expenses *preparing, processing, administering and coordinating exams.* *Hint: Recognize one.* Not for teaching or training.

The *VECs* maintain the pools of questions. The Examiner Coordinator organizations maintain the pools, not the individual examiners.

COMMISSION'S RULES

The Volunteer Examiner Coordinator is an *organization that has entered into an agreement with the FCC* **to coordinate, prepare and administer amateur radio license examinations.** *Hint: The coordinator is an organization that coordinates.*

To be accredited as a Volunteer Examiner a *VEC must confirm the VE applicant meets FCC requirements.* *Hint: A VE is accredited by a VEC, not the FCC.*

If the examinee does not pass the exam, the VE team will *return the application form* **to the examinee.** You get the application back, not the fee.

Each administering VE **is responsible for the proper conduct and necessary supervision during an amateur radio license examination session.** Every individual is responsible for maintaining the integrity of the system. We don't want the FCC to take examinations back in-house.

If a candidate fails to comply with the examiner's instructions, the examiner should *immediately terminate the candidate's exam.* Not everyone's, just the one.

A VE may not administer an examination to someone who is a *relative of the VE* **as listed in the FCC rules.** *Hint: Friends and employees are OK.*

A VE who fraudulently administers or certifies an examination faces *revocation of his amateur station license* **and suspension of his amateur license.** No fine or jail time – loss of license is worse!

After administering a successful examination, the VEs must *submit the application to the coordinating VEC* **according to the VEC's**

instructions. The test is graded on-site by the VEs who send the results to their VEC. The VEC loads the data in the FCC database.

When an examinee scores a passing grade, *three VEs must certify* that the examinee is qualified for the license grant and that they have complied with the VE requirements. *Hint: Look for "three VEs," and that is the answer.*

E1F Miscellaneous Rules

Spread-spectrum transmissions change frequency in a deliberate pattern to decrease interference and assure privacy. The result is a wide signal, so it is limited to the higher and wider bands. **Spread-spectrum transmissions are permitted only on amateur frequencies *above 222 MHz.***

A Canadian license holder in the US is *allowed his Canadian privileges*, not to exceed the US Amateur Extra privileges. *Hint: Not to exceed US privileges.*

Amplifiers must be FCC certified to protect against use by CBers and to assure spectrum purity. **A dealer may sell an external RF power amplifier capable of operation below 144 MHz if it has not been granted FCC certification if it *was constructed or modified by an amateur operator for use at an amateur station.*** *Hint: Too complicated! Just remember, amateurs can construct and modify.*

Line A is a line roughly parallel to and south of the *US-Canadian border.*

Amateur stations in the US may not transmit on *420 MHz – 430 MHz* if they are north of Line A.

COMMISSION'S RULES

A Special Temporary Authority may be issued by the FCC to provide for *experimental amateur communications.* These usually allow for operation outside normal amateur frequencies. This is *not* a special event station.

An amateur station may send a message to a business only when neither the amateur nor his employer has any *pecuniary interest* in the communications. I can send congratulations to a business but not a product order.

Amateur radio isn't for business. ***Communications transmitted for hire or material compensation, except as otherwise provided in the rules, are prohibited.*** *Hint: "For hire" is prohibited. You can't be paid for being a Ham.*

An amateur radio mesh network may not transmit *messages encoded to obscure their meaning.* *Hint" That's easy. No one can.*

An auxiliary station is one that controls another station over a radio link. **The control operator of an auxiliary station may be a *Technician*, General, Advanced, or Amateur Extra operator.** Advanced and Novice are old license classes. *Hint: Look for the only answer that includes "Technician."*

To qualify for a grant of FCC certification, an amplifier must *satisfy the FCC's spurious emissions standards* when operated at 1500 watts or its full output power
Hint: Satisfy the FCC for certification. Don't bother memorizing the power.

SUMMARY: COMMISSION'S RULES
GROUP 1A – FREQUENCY PRIVILEGES
Don't allow your signal over the band edge.

3 kHz USB must be 3 kHz below band edge.

3 kHz LSB must be 3 kHz above band edge.

14.350 MHz AND 3.6 MHz.

Control operator on US vessel must hold FCC license or be authorized for alien reciprocal operation.

On 60 meters set carrier frequency in the center of the channel.

Max power on 2200 meters is 1 watt.

Max power on 630 meters is 5 watts.

Control operator of originating station is responsible.

Must get permission from master or pilot to operate.

Any amateur license in international waters.

GROUP 1B – STATION RESTRICTIONS
Spurious emission is outside necessary bandwidth.

Slow scan TV is 3 kHz.

Protect FCC monitoring facility within 1 mile.

Cease operation or make changes if interfering with a radiolocation system.

National Radio Quiet Zone is around National Radio Astronomy Observatory.

Notify FAA before installing antenna near airport.

PRB-1 applies to state and local zoning.

FCC may place limits to avoid interference.

RACES certifies members.

RACES rules allow all frequencies authorized to the control operator.

GROUP 1C – CONTROL
Max bandwidth for data is 2.8 kHz.

Communicate remarks of a personal nature.

Wait 30 days after notifying UTC of intended operations.

COMMISSION'S RULES

Notify UTC of your call sign and coordinates.

IARP allows ops in the Americas.

CEPT allows ops in Europe.

CEPT requires you have Public Notice DA 16-1048.

Automatic control third-party only with RTTY or data.

Remote control must time-out after 3 minutes.

Highest modulation index below 29 MHz is 1.

Permitted spurious is -43 dB below the fundamental.

Phone permitted in entire 630-meter band.

GROUP 1D –SPACE AND EARTH STATIONS

Telemetry is one-way transmission of measurements.

Telecommand may be encrypted.

Telecommand commands function of a space station.

Balloon station must identify with a call sign.

Station operated by telecommand must post copy of license, name address and phone number.

Model aircraft telecommand limited to 1 watt.

HF space stations on 40, 20, 15 and 10 meters.

VHF only 2 meters.

UHF only 70 cm and 13 cm.

Must be designated by the space station licensee to telecommand.

Any amateur station can operate as an Earth station subject to the privileges of the control operator.

One-way communications permitted from a space station, beacon, or telecommand.

GROUP 1E – VOLUNTEER EXAMINER PROGRAM

Reimbursed for preparing, processing, administering exams.

VECs maintain the question pool.

Volunteer Examiner Coordinator coordinates, prepares, and administers.

To be a VE, the VEC must confirm you meet requirements.

If you don't pass, VE team returns your application.

Each VE is responsible for proper conduct.

If applicant doesn't follow instructions, examiner should immediately terminate his exam.

VE may not administer to a relative.

Fraudulent VE faces revocation and suspension.

VEs submit a successful application to the VEC.

Three VEs must certify passing.

GROUP 1F – MISCELLANEOUS RULES

Spread spectrum only above 222 MHz.

Canadian allowed his privileges.

May sell amplifier not certified by the FCC if it was constructed or modified by an amateur radio operator for use at an amateur station.

Line A is parallel and south of US-Canada border.

May not transmit on 420–430 MHz north of Line A.

Special Temporary Authority is for experimental.

May send a message to a business only if no pecuniary interest.

Communications for compensation are prohibited.

Mesh network may not obscure meaning.

Control operator of an auxiliary station may be a Technician.

Amplifier must satisfy FCC spurious emissions standards.

OPERATING PROCEDURES

E2A Amateur Radio in Space

The ascending pass for an amateur satellite is from south to north. *Cheat: "Ascending." Think of it as going up from the South Pole to the North.*

A transponder receives a signal and emits a signal in response. A satellite might receive a signal on one band (uplink) and retransmit on another (downlink).

When a satellite is using an inverting linear transponder:
- **Doppler shift is reduced because the uplink and downlink shifts are in opposite directions**
- **Signal position in the band is reversed**
- **Upper sideband on the uplink becomes lower sideband on the downlink, and vice versa.**
- **All these choices are correct.**

Hint: Everything in the answers is inverted.

An upload signal is processed by an inverting linear transponder as the signal is *mixed with a local oscillator signal* and the difference product is transmitted. *Hint: Processed by mixing.*

A satellite's mode is the *uplink and downlink frequency bands.*

The letters in a satellite's mode designator specify the *uplink and downlink frequency ranges.*
If a satellite is operating in U/V mode, it means the uplink is UHF, and the downlink is VHF.

Keplerian elements are parameters that define the *orbit of a satellite*. Hint: Kepler developed formulas for describing an orbiting body.

A linear transponder can relay:
- **FM and CW**
- **SSB and SSTV**
- **PSK and Packet**
- **All these choices are correct**

Hint: The transponder is linear, so it can relay any type of signal.

The effective radiated power to a satellite that uses a linear transponder should be limited *to avoid reducing the downlink power* to all other users. A satellite receiving a very strong signal assumes an excellent connection and will reduce its transmit power to conserve energy.

The terms "L band" and "S band" refer to the *23-centimeter and 13-centimeter bands*. Cheat: Remember 23 or 13, and you have the answer.

A satellite that stays in one position in the sky is *geostationary*. Hint: If it stays in one place, it is stationary.

To minimize the effect of spin modulation and Faraday rotation, use a *circularly polarized antenna*. Cheat: Spin in circles.

The purpose of digital store and forward functions is to *hold digital messages in the satellite for later download*.
Hint: Digital store and forward stores digital messages and forwards them later.

Digital satellites relay messages by *store-and-forward*. Hint: They store the message until they get over the relay target area.

E2B Television Practices

In digital television, a coding rate of 3/4 means **25% of the data sent is forward error correction data.** *Hint: 3/4 of the data is real, 1/4 is error correction.*

Fast-scan (NTSC) television transmits **525** lines per frame.

An interlaced scanning pattern is generated in a fast-scan (NTSC) television by scanning *odd numbered lines in one field and even numbered lines in the next.* *Hint: Odd and even are interlaced.*

Color information in analog SSTV is sent by color lines *sent sequentially* — one color after the other. SSTV is slow-scan, used mainly to send pictures.

Vestigial sideband is used to *reduce bandwidth while increasing the fidelity of low frequency video.* *Hint: Vestigial means a small remnant, reducing bandwidth.*

Vestigial sideband modulation is amplitude modulation in which one complete sideband and *a portion of the other are transmitted.* *Hint: Vestigial means a small remnant – a portion of the other sideband.*

The modulation used for amateur television DVB-T signals is *QAM and QPSK.* DVB-T is terrestrial digital television. QAM and QPSK are modulation methods. *Hint: Remember the Qs.*

TV receivers can be used for fast-scan TV operations on the 70-cm band by *transmitting on channels shared with cable TV.* *Hint: To use a TV receiver, transmit on a channel it receives.*

To receive and decode SSTV using Digital Radio Mondiale (DRM), use an *SSB receiver*.

The brightness of an analog slow-scan picture is encoded by a *tone frequency*.

SSTV receiving software is signaled to begin a new line by *specific tone frequencies*.
Hint: Look for "tone frequencies" in both answers.

The Vertical Interval Signaling (VIS) code is sent as part of an SSTV transmission to *identify the SSTV mode* being used. *Hint: It is signaling the mode.*

E2C Contest and DX Operating

When a US licensed operator is operating a remote control transmitter located in the US, *no additional indicator is required* (when identifying). The transmitter is in the US and US rules apply.

Amateur radio log data is exchanged using *ADIF file format.* Amateur Data Interface Format.

Amateur contesting is generally excluded on *30 meters.* Contesting is not allowed on any of the WARC bands (30, 17, and 12 meters).

A mesh network is interconnected wireless points using a wireless network adapter. **The frequencies that can be used for mesh networks are *shared* with various unlicensed wireless data services.** *Hint: Wireless network adapters are unlicensed.*

A DX QSL manager handles the *receiving and sending* of confirmation cards for a DX station.

During a VHF/UHF contest, you would expect to find the highest level of activity in the *weak*

OPERATING PROCEDURES

signal segment **of the band with most of the activity near the calling frequency.** *Hint: Most people will listen for weak signals near the weak-signal calling frequency.*

Cabrillo format is a standard for submitting *electronic contest logs.* ADIF is used for data interchange among programs. Cabrillo is for reporting.

Contacts which may be confirmed through Logbook of the World (LoTW) are
- **Special event contacts between stations in the US**
- **Contacts between US and non-US stations**
- **Contacts for Worked All States credit**
- **All these choices are correct**

LoTW is used for all types of contacts.

The equipment commonly used to implement a ham radio mesh network is a *wireless router* **running custom software.**

A DX station might transmit and receive on different frequencies:
- **Because they are transmitting on a frequency prohibited to some of the responding stations.**
- **To separate the calling stations from the DX station.**
- **To improve the operating efficiency by reducing interference.**
- **All these choices are correct.**

Many DX stations operate "split," listening on a different frequency from their transmissions. Get my book for details, "How to Chase, Work and Confirm DX. – The Easy Way."

When attempting to contact a DX station during a contest or pileup, generally identify by sending

your full call once or twice. Hint: "Full call." No partial calls, no grid squares, no repetitive identifying.

The term used to describe a delay between a control operator action and the corresponding change in the transmitted signal is called "*Latency.*"

E2D Operating Methods: VHF/UHF

The digital mode designed for meteor scatter is *MSK144.* Hint: Meteor **SK**atter

The information replacing signal-to-noise ratio when using FT8 or FT4 modes in a VHF contest is *Grid Square.* Hint: Collecting grid squares is popular on VHF.

A digital mode designed for EME communications is *Q65.* EME is Earth-Moon-Earth; a slow mode, like Q65, designed for extremely weak signals. Hint: If you see" Q65," that is the answer.

The technology used for real-time tracking balloons carrying amateur radio transmitters is *APRS.* APRS is Automatic Packet Reporting System. A GPS interfaces with a radio to send and receive position reports.

A characteristic of JT65 mode is it *decodes signals with a very low signal-to-noise ratio.* Hint: JT65 decodes signals below the noise.

The method of establishing EME contacts is *time-synchronous* transmissions alternating between stations. Hint: To establish contact, you need to know when to listen (time synchronous).

The digital protocol used by APRS is *AX.25.* Hint: If you see AX in an answer, it is correct.

The type of packet frame used to transmit APRS beacon data is *unnumbered information.*   Hint: *Data is information.*

The type of modulation used for JT65 contacts is *multi-tone AFSK.* AFSK is Audio Frequency Shift Keying. *Hint: JT65 uses audio tones to decode signals with a very low signal-to-noise ratio.*

A packet path WIDE3-1 designates *three digipeater hops* are requested with one remaining. A digipeater is a repeater for digital signals. Packet is digital. *Hint: A packet path would go through a digipeater.*

An APRS station relays data by packet *digipeaters.* Hint: *Digipeaters relay data.*

E2E Operating Methods: HF digital

A common type of modulation for data emissions below 30 MHz is *FSK.* Frequency Shift Keying is used for RTTY (teletype). FSK is the only answer that is a digital (data) mode. *Hint: Recognize that FSK is a digital mode.*

WSJT-X digital mode transmit/receive timing is synchronized by *synchronization* of computer clocks. *Hint: Synchronize by synchronizing.*

The "4" in FT4 refers to *four-tone* continuous-phase frequency shift keying. FT4 is a data mode. *Cheat: Look for the answer with "continuous-phase."*

What is characteristic of the FTS4 mode?
- **Four-tone Gaussian frequency shift keying**
- **Variable transmit/receive periods**
- **Seven different tone spacings**
- **All these choices are correct**

The digital mode that does NOT support keyboard-to-keyboard operation is *WSPR*.
Cheat: Whisper, this is not a chat room.

The length of an FT8 transmission cycle is *15 seconds*.

Q65 differs from JT65 in that *multiple receive cycles* are averaged.

The HF digital mode used to transfer binary files is *PACTOR*. PACTOR sends and receives digital information using radio.

***PSK31* uses variable-length character coding.**
The letters have different lengths. Capital letters take twice as long to send, so don't use all caps.

Of the modes listed, *FT8* has the narrowest bandwidth (50hz).

The difference between direct FSK and audio FSK is direct FSK *modulates the transmitter VFO*.
FSK is Frequency Shift Keying. Diddling the VFO changes the tone on the receiving end. Audio FSK feeds an audio tone to the transmitter, sending SSB.
ALE stations establish contact by *constantly scanning* a list of frequencies, activating the radio when the designated call sign is received.
ALE is Automatic Link Enable. The radio scans and automatically establishes contact.

The digital mode with the fastest data throughput under clear conditions is PACTOR IV.
PAcket Teleprinting Over Radio uses multiple channels to increase speed.

SUMMARY: OPERATING PROCEDURES

GROUP 2A AMATEUR RADIO IN SPACE

Ascending pass is South to North.

Inverting linear transponder reduces Doppler, reverses signals and sidebands.

Inverting linear transponder processes by mixing with local oscillator.

Satellite's mode is uplink and downlink frequency bands.

Letters in mode designator specify the uplink and downlink frequency ranges.

Keplerian elements define orbit of a satellite.

Linear transponder can relay all modes.

Reduce power to a satellite to avoid reducing the downlink power.

L band and S band refer to 23-cm and 13-cm bands.

A satellite that stays in one position is geostationary.

To minimize effects of spin modulation and Faraday rotation, use circularly polarized antenna.

Store and forward holds messages for later download.

Relay messages by store and forward.

GROUP 2B – TELEVISION

Coding rate of 3/4 means 25% is forward correction data.

Fast scan TV transmits 525 lines per frame.

Interlaced pattern scans odd and even lines.

Color sent by color lines sequentially.

Vestigial sideband reduces bandwidth.

DVB-T signals are QAM abd QPSK.

TV receivers can be used by transmitting on cable channels.

SSTV using Digital Radio Mondiale use a SSB receiver.

SSTV begins a new line with specific tone frequencies.

Vertical Interval signaling code identifies the SSTV mode being used.

GROUP 2C – CONTEST AND DX OPERATING

No additional indicator required when operating remote in the US

Log data exchanged by ADIF file format.

Contesting excluded on 30 meters.

Mesh networks share frequencies with unlicensed data services.

QSL manager handles receiving and sending confirmation cards.

Most activity is in the weak signal segment.

Cabrillo format is used for submitting electronic logs.

LoTW can confirm all contacts.

Mesh network uses wireless router and custom software.

DX may receive and transmit on different frequencies for all the reasons.

Identify by sending your full call once or twice.

Delay is called "latency."

GROUP 2D – VHF /UHF OPERATING

Meteor scatter uses MFK144.

Contesting FT8 send grid square.

EME uses Q65.

Real-time tracking uses APRS.

JT65 decodes signals with low signal-to-noise ratio.

EME contacts use time-synchronous transmissions.

APRS uses AX.25.

APRS beacon data is unnumbered information.

JT65 uses multi-tone AFSK.

WIDE3-1 is 3 digipeater hops.

APRS relays data by packet digipeaters.

GROUP 2E – HF OPERATING

Data modulation below 30 MHz is FSK.

OPERATING PROCEDURES

WSJT timing uses synchronized computer clocks.

FT4 uses four-tone continuous phase frequency shift.

FTS4 uses all the choices.

WSPR does not support keyboard-to-keyboard.

FT8 transmission cycle is 15 seconds.

Q65 averages multiple receive cycles.

HF digital mode for binary files is PACTOR.

PSK31 uses variable-length character spacing.

FT8 has the narrowest bandwidth.

Direct FSK modulates the transmitter VFO.

ALE stations constantly scan.

Fastest data throughput is PACTOR IV.

E3 – RADIO WAVE PROPAGATION

E3A *Electromagnetic Waves*

The maximum separation between two stations communicating by EME is 12,000 miles, if the Moon is "*visible*" by both stations. EME is moon bounce, Earth-Moon-Earth. *Hint: Forget the miles. The only way Earth-Moon-Earth works is if the Moon is visible by both stations. Common sense.*

Libration fading of an EME signal is a fluttery irregular *fading*. Libration is a perceived oscillating motion of two orbiting bodies. *Hint: Fading is fading.*

When scheduling an EME contact, the condition of least path loss is when the Moon is at *perigee*. Perigee is nearest to the Earth, so your signal travels a shorter distance. Apogee is afar.

An electromagnetic wave travels at a *right angle* to the electric and magnetic fields. We usually say the fields are at right angles to the wave.

The component fields of an electromagnetic wave are oriented at *right angles*. Electric and magnetic fields are at right angles.

If the MUF decreases due to darkness, *switch to a lower frequency* HF band. MUF is the maximum usable frequency. If it decreases, follow it down. The lower frequencies come to life after dark.

Atmospheric ducts capable of propagating microwave signals often form over *large bodies of water*.

RADIO WAVE PROPAGATION

When a meteor strikes the Earth's atmosphere, a linear ionized region is formed in the **E region**.
Cheat: MEteor, Earth, linEar, E Region.

Meteor scatter is most suited to **28 MHz – 148 MHz.** *Hint: Six meters is popular for meteor scatter, and this answer is the only one that includes 50 MHz (six meters). "Meteor" has six letters, and that helps me remember "six meters."*

The speed of electromagnetic waves through a medium is determined by the **index of refraction**.
The wave refracts (bends).

The typical range for tropospheric duct propagation is **100 – 300 miles.**

Severe geomagnetic storms result in auroral propagation. Auroral activity is caused by charged particles from the Sun interacting with the Earth's magnetic field. Geomagnetic storms produce charged particles.

The best mode for auroral propagation is **CW.**
Hint: CW has a 10 dB advantage over SSB and is always the best mode for propagation compared to the other answers.

Circularly polarized electromagnetic waves are waves with **rotating electric and magnetic fields.**
Hint: Circles rotate.

E3B Transequatorial Propagation

Transequatorial propagation (TEP) is most likely to occur between points separated by 2,000 to 3,000 miles over a path **perpendicular to the geomagnetic equator.**
Hint: Forget the distance. If it is transequatorial, it must be perpendicular to the equator.

The approximate maximum range for transequatorial propagation is **5,000 miles.**

The best time for transequatorial propagation is **afternoon or early evening.** *Hint: Give the sun all day to charge up those electrons.*

Extraordinary and ordinary waves are **independently propagating, elliptically polarized waves created in the ionosphere.** *Cheat: Extraordinary and ordinary are independent. Look for "independent."*

Long-distance propagation on 160 meters is most likely over a **path entirely in darkness.** 160 is a night-time band.

Long-path propagation is most frequent on **40 meters and 20 meters.** *Hint: The band usually open for all DX is 20 meters.*

The effect of lowering a signal's transmitted elevation angle for ionospheric HF skip propagation is **the distance covered by each hop increases.** The lower angle travels further before refracting off the ionosphere and refracts at a lower angle, traveling further before returning to Earth.

When the signal frequency is increased, the maximum range of ground-wave propagation **decreases.** Higher frequencies don't travel as well by ground wave.

The time of year sporadic E propagation is most likely to occur is around the solstices, especially the **summer solstice.** Sporadic E is a summer phenomenon.

Sporadic E occurs mostly likely **between sunrise and sunset.** It is a day-time occurrence.

RADIO WAVE PROPAGATION

The effect of chordal hop propagation is that the signal experiences less loss than multi-hop propagation, which *uses the Earth as a reflector.* Chordal hop means the signal reflects inside the ionosphere and doesn't hit the Earth until it finally comes back down at the receiver. *Hint: Less loss is a good thing.*

Chordal hop is successive ionospheric reflections *without an intermediate reflection from the ground.*

The polarization supported by ground-wave propagation is *vertical.* Vertical polarization spreads the signal along the ground.

E3C Radio Propagation

The cause of short-term radio blackouts is *solar flares.*

A rising A or K index indicates *increasing disturbance* of the geomagnetic field. *Hint: The index rises with geomagnetic disturbances.*

When the A or K index is elevated, the paths most likely to experience high levels of absorption are through the *auroral oval.* *Hint: An elevated A or K index indicates a geomagnetic disturbance, and that would be strongest around the poles, the auroral oval.*

The value of Bz (B sub z) represents the *north-south strength* of the interplanetary magnetic field. *Hint: B and z represent two things, direction and strength. The other answers are only one thing.*

The orientation of Bz that increases the likelihood charged particles from the Sun will cause disturbed conditions is *southward.* *Hint:*

If the waves are headed southward (from the north), they will hit the North magnetic pole and cause disruptions.

You learned on the Technician Class test that UHF/VHF radio waves can bend slightly over the horizon. **The radio horizon, compared to the geographic horizon, is *approximately 15% farther*.** *Cheat: Farther but not by a lot, and this is the lowest answer.*

Solar flares are categorized by a letter. **The greatest solar flare intensity is *Class X*.** *Hint: X as in "eXtreme."*

Space weather also gets a letter. **The term *G5* means an extreme geomagnetic storm.** *Hint: "G" as in "geomagnetic."*

Data reported by amateur radio propagation reporting networks is *digital-mode and CW* signals.

The 304A solar parameter measures UV emissions at *304 angstroms*, correlated to the solar flux index. *Cheat: 304 A as in "Angstroms."*

VOCAP software models *HF propagation*. Voice of America Coverage Analysis Program.

A sudden rise in radio background noise across a large portion of the HF spectrum indicates a *coronal mass ejection impact or solar flare* has occurred.

SUMMARY: RADIO WAVE PROPAGATION

GROUP 3A – ELECTROMAGNETIC WAVES

For EME moon must be visible to both stations.

Libration fading is fluttery, irregular.

EME best when moon at perigee.

Electromagnetic wave travels at right angles to the electric and magnetic fields.

Fields in electronic wave are at right angles.

When MUF decreases move to a lower frequency.

Atmospheric ducts form over large bodies of water.

Meteor ionizes the E region.

Meteor scatter 28 MHz – 148 MHz.

Speed of waves determined by index of refraction.

Tropospheric duct 100- 300 miles.

Geomagnetic storms result in auroral propagation.

Best mode for auroral propagation is CW.

Circularly polarized waves have rotating electric and magnetic fields.

GROUP 3B – TRANSEQUITORIAL PROPAGATION

Most likely perpendicular to the equator.

Range is 5,000 miles.

Best in afternoon to early evening.

Extraordinary and ordinary waves are independently propagating.

Long distance on 160 meters in darkness.

Long-path mostly on 40 and 20 meters.

Lower elevation angle increases distance covered by each hop.

Ground-wave range decreases at higher frequencies.

Sporadic E mostly around summer solstice.

Sporadic E between sunrise and sunset.

Chordal hop has less loss.

Ground-wave propagation is vertically polarized.

GROUP 3C – RADIO PROPAGATION

Short-term radio blackouts caused by solar flares.

Rising A or K index indicates increasing geomagnetic disturbance.

High A or K index means higher absorption on paths through the auroral hole.

Bz is north-south magnetic field strength.

Increased disturbed conditions with Bz southward.

Radio horizon is 15% further than geographic.

Greatest solar flare is Class X.

G5 is extreme geomagnetic storm.

Networks report digital-mode and CW signals.

304A measures at 304 angstroms.

VOCAP software models HF propagation.

Sudden rise in noise indicates coronal mass ejection or solar flare.

E4 – AMATEUR PRACTICES

E4A Test Equipment

The highest frequency signal that can be accurately displayed on a digital oscilloscope is limited by the *sampling rate* of the analog-to-digital converter. *Hint: The frequency is limited by the rate the oscilloscope can sample: a higher sampling rate = higher frequency.*

A spectrum analyzer would display *signal amplitude and frequency*. *Hint: It is analyzing the spectrum and shows the frequency and strength of the signal.*

To display spurious signals and/or intermodulation distortion products in an SSB transmitter, use a *spectrum analyzer*. *Hint: Analyze the spectrum to spot spurious signals.*

Calibrating an oscilloscope probe is called "compensating." **Compensation of an oscilloscope probe is typically done by displaying a *square wave* and adjusting the probe so the horizontal lines are as flat as possible.** *Hint: A square wave display shows symmetrical flat lines.*

A prescaler on a frequency counter *reduces the signal frequency* to within the counter's operating range. *Hint: It scales the signal before it is measured.*

The effect of aliasing on a digital oscilloscope is that a *false, jittery* low-frequency version of the signal is displayed. *Hint: Forget why it happens. Aliasing is another word for distortion, false display.*

The advantage of an antenna analyzer over an SWR bridge is antenna analyzers *compute SWR and impedance* automatically. *Hint: An analyzer computes it all, not just SWR.*

The following are used to measure SWR
- Directional wattmeter
- Vector network analyzer
- Antenna analyzer
- All these choices are correct

When using an oscilloscope probe, it is good practice to *minimize the length of the probe's ground connection.* *Hint: A short ground lead has less effect on the measured circuit.*

The most effective trigger mode when using an oscilloscope to measure a linear power supply's output ripple is *"line."* *Cheat: Line is linear.*

An antenna analyzer can measure
- Velocity factor
- Cable length
- Resonant frequency of a tuned circuit
- All these choices are correct

E4B Measurements

The accuracy of a frequency counter is most affected by the *time base accuracy*. *Hint: The base must be accurate.*

The significance of voltmeter sensitivity expressed in ohms per volt is a full-scale reading of the voltmeter multiplied by its ohms per volt is the *input impedance*. *Hint: Algebra. Volts x ohms/volt = ohms. Ohms indicate impedance. Cheat: The word "voltmeter" is in both the question and answer.*

AMATEUR PRACTICES

The subscripts of S parameters represent the *port or ports at which measurements are made.* *Cheat[3]: S as in "portS."*

The S parameter equivalent to forward gain is *S21.* *Cheat: We looked forward to turning 21.*

The S parameter that represents return loss or SWR is *S11.* *Cheat: A 1:1 SWR is good.*

To calibrate a RF vector network analyzer, you use a *short circuit, open circuit, and 50-ohm loads.* *Hint: You calibrate at the two extremes and the "normal" load of 50-ohm coax.*

The power absorbed by the load when the forward power is 100 watts and the reflected power is 25 watts is *75 watts.* *Hint: Simple math. 100 watts went out, and 25 came back, so 75 watts must have stayed in the load.*

The Q for a series-tuned circuit can be determined by the *bandwidth* **of the circuit's frequency response.** Higher "Q" means narrower bandwidth. *Hint: You measure Q by measuring the bandwidth. Forget about "series tuned."*

A two-port vector network analyzer can measure *filter frequency response.*

To measure intermodulation distortion (IMD) in an SSB transmitter, modulate with two non-harmonically related *AF signals* **and observe the output on a** *spectrum analyzer.* *Hint: You*

---

[3] There are random questions that mean nothing to you. The only way to recognize the correct answer is with a cheat. These cheats have nothing to do with a correct answer but will help you recognize one.

*modulate with AF (audio), and you measure distortion
with a spectrum analyzer.*

A vector network analyzer, can measure:
- **Input impedance**
- **Output impedance**
- **Reflection coefficient**
- **All these choices are correct**

*It analyzes the network parameters of an electrical
circuit: in, out, and reflection.*

E4C Receiver Performance

**An effect of excessive phase noise in an SDR
receiver's master clock oscillator is it can
*combine with strong signals on nearby
frequencies to generate interference.*** *Hint:
Excessive noise causes interference.*

**The receiver circuits that can be effective in
eliminating interference from strong out-of-band
signals are a *front-end filter or pre-selector.***
*Hint: The signals are out-of-band and need to be
filtered or pre-selected at the front end before they
get into the more sensitive internals of the receiver.*

**The term for the suppression in an FM receiver
by another stronger signal on the same
frequency is *capture effect.*** *Hint: FM receivers
capture the strongest signal.*

**The noise figure of a receiver is defined as the
ratio in dB of the noise generated by the receiver
to the *theoretical minimum noise.*** *Hint: Noise
figure is receiver-generated noise measured against a
theoretical minimum.*

**The value of -174 dBm/Hz noise floor represents
the theoretical noise at the input of a perfect**

receiver *at room temperature*. *Cheat: The only answer on the test mentioning "room temperature."*

Increasing a receiver's bandwidth from 50 Hz to 1,000 Hz increases the receiver's noise floor by 13 dB. Opening the bandwidth allows in more noise.

The MDS of a receiver is the *minimum discernible signal*. It is a measure of receiver sensitivity.

An SDR is a software-defined radio. An analog-to-digital converter is the heart of an SDR. It converts the signals to digital ones and zeros. Then, computer power sorts them out. **An SDR receiver is overloaded when signals exceed the *reference voltage* of the analog-to-digital converter.** *Hint: Signals exceeding the reference voltage overload the receiver.*

Conventional super-heterodyne receivers convert the incoming signal to a fixed intermediate frequency (IF) and feed it through filters designed for that frequency. **A good reason for selecting a high frequency for the IF is it is easier for the front end circuitry to *eliminate image responses*.** Mixing "up" to a higher frequency puts the images farther away from the intermediate frequency.

The advantage of having a variety of receiver IF bandwidths to choose is that *receiver bandwidth can be set to match the modulation bandwidth*, maximizing signal-to-noise ratio and minimizing interference. *Hint: That is way too long an answer. Remember to match the receiver to the mode.* You can use a narrow filter for CW and a wider one for SSB.

An attenuator reduces receiver overload on the lower HF bands with little or no impact on the signal-to-noise ratio because *atmospheric noise*

is greater than internally generated noise. An attenuator reduces external noise. Reduce external noise to the noise level generated in the receiver so the signal stands out.

A narrow-band roofing filter improves blocking dynamic range by *attenuating strong signals near the receive frequency*. *Hint: A filter attenuates signals near the receive frequency.*

Reciprocal mixing is *local oscillator phase noise* mixing with adjacent strong signals to create interference. *Hint: Remember, local oscillator phase noise causes reciprocal mixing.*

The purpose of a receiver IF Shift control is to *reduce interference* from stations transmitting on adjacent frequencies. Shift the Intermediate Frequency to move the signal within a filter's bandwidth.

E4D Receiver Performance Characteristics

The blocking dynamic range of a receiver is the difference in dB between the noise floor and the level of an incoming signal that will cause *1 dB of gain compression*. *Hint: Blocking dynamic range is "gain compression."*

Poor dynamic range in a receiver are described as spurious signals caused by cross modulation and *desensitization* from strong adjacent signals. *Hint: Strong adjacent signals desensitize a receiver with poor dynamic range. Forget the rest of the answer.*

Intermodulation interference can occur between two repeaters in close proximity when the *output signals mix* in the final amplifier of one or

both transmitters. *Hint: Modulation and intermodulation are mixing.*

To reduce or eliminate intermodulation interference in a repeater, use a *properly terminated circulator* at the output of the transmitter. *Hint: To eliminate, terminate.* A circulator is a one-way valve that shunts off signals coming down the transmission line and sends them to a dummy load (properly terminated).

If a receiver is tuned to 146.70 MHz and a nearby station transmits on 146.520 MHz, the transmitter frequencies which would produce an intermodulation-product signal in the receiver are 146.34 MHz and 146.61 MHz. *Hint/Cheat: Take the average of the two (146.70 + 146.52)/2 and look for the answer that has 146.61. You'll only solve one, which is enough to choose the correct answer.*

The term for the reduction in receiver sensitivity caused by a strong signal near the received frequency is *desensitization*. *Hint: reduction in sensitivity is desensitization.*

To reduce desensitization, *insert attenuation*. *Hint: Reduce the strong signal by adding attenuation.*

Intermodulation in electronic circuits is caused by *nonlinear circuits or devices*.

The purpose of a preselector in a receiver is to increase *rejection of signals outside the desired band*. *Hint: It preselects signals.*

A third-order intercept level of 40 dBm means a pair of 40 dBm signals will generate a third-order intermodulation *product with the same output amplitude as the input signals*. *Cheat: A pair are the same.*

Odd-order intermodulation products are of particular interest because the odd-order product of two signals which are in a band of interest will also *likely be within the band.*
Hint: You are interested because the products will be in the band you are working in.

Link margin refers to the difference between the received signal strength (dB) and the minimum signal the receiver can hear. **The link margin in a system with a transmit power level of 10 W (+40 dBm), a system antenna gain of 10 dBi, a cable loss of 3 dBi, a path loss of 136 dB, a receiver minimum discernable signal of -103 dBm and a required signal-to-noise ratio of 6 dB is +8 dB.** *Solve: First, the signal strength is 40+10-3-136 = -89 dB. The receiver can hear down to -103 but needs 6 dB signal-to-noise ratio = -97. The difference is + 8dB. Cheat: The link margin must be positive to hear anything. There is only one positive answer.*

The received signal level with a transmit power of 10 W (+40 dBm), a transmit antenna gain of 6 dBi, a receive antenna gain of 3 dBi, and a path loss of 100 dB is -51 dBm.
Solve: 40+6+3-100 = -51.

A minimum discernible signal of -100 dBm represents 0.1 *picowatts.* dBm is decibels in reference to a milliwatt (.001 watts). -100 dB is one ten-millionth of that. *Cheat: Pick the lowest answer.*

E4E Noise and Interference

A problem when using automatic notch-filtering to remove interfering carriers while receiving CW signals is it *removes the CW signal* as well as the interfering carrier.
Hint: The automatic filter removes all tones, including the CW tone.

Noise that can be reduced with digital noise reduction includes:
- **Broadband white noise**
- **Ignition noise**
- **Power line noise**
- **All these choices are correct**

Hint: Digital signal processing is effective against all sorts of noise.

A noise blanker removes *impulse noise*. A noise blanker is effective against pops from electric fences and spark plugs.

Conducted noise from an automobile battery charging system can be suppressed by installing *ferrite chokes* on the charging system leads.
Hint: Choke off noise conducted in the lines.

Radio frequency interference from a line-driven AC motor can be suppressed by installing a *brute-force* AC-line filter in series with the motor leads. *Hint: Suppress noise with brute-force.*

Computer network equipment might cause *unstable modulated or unmodulated signals at specific frequencies*. *Hint: Computer circuits can generate "birdies" – signals heard at specific frequencies.*

Shielded cables can radiate or receive interference due to *common-mode currents* on

the shield and conductors. *Hint: The interference is common to the shield and conductors.* Solve that by putting ferrite chokes on the cable.

Current that flows equally on all conductors is called *common-mode* current. *Hint: It is common to all the conductors.*

An undesirable effect when using a noise blanker is *strong signals may be distorted and appear to cause spurious emissions.* *Hint: Signal processing can cause distortion.*

Intermittent loud roaring or buzzing AC line interference could be:
- Arcing contacts in a thermostatically controlled device.
- Defective doorbell or doorbell transformer.
- Malfunctioning illuminated advertising display.
- All these choices are correct.

Hint: All operate intermittently.

Local AM broadcast signals can combine to generate spurious signals on the MF or HF bands, most likely is caused by nearby *corroded metal connections* mixing and re-radiating the broadcast signals. Corroded metal joints can act as diodes and mix signals.

Interference received as a series of carriers at regular intervals across a wide frequency range is caused by *switch-mode power supplies*.

An AC surge suppressor should be installed on the *single point ground panel*.

The purpose of a single point ground panel is to ensure all lightning protectors *activate at the same time.* Everything at the same potential.

SUMMARY: AMATEUR PRACTICES

GROUP 4A – TEST EQUIPMENT

Highest frequency accurately displayed on a digital oscilloscope determined by the sample rate.

Spectrum analyzer shows amplitude and frequency.

Display spurious signals on a spectrum analyzer.

Compensation with a square wave.

Prescaler reduces frequency to within the counter's range.

Aliasing effect is false, jittery, low-frequency version.

Antenna analyzer computes SWR and impedance automatically.

SWR measured with directional wattmeter, vector network analyzer and antenna analyzer.

Minimize length of oscilloscope's ground connection.

To measure output ripple use trigger mode is "line."

Antenna analyzer can measure velocity factor, cable length and resonant frequency.

GROUP 4B – MEASUREMENTS

Accuracy of frequency counter most affected by time base accuracy.

Full-scale reading times ohms per volt rating equals input impedance.

Subscript of "S" parameters are measurement ports.

S21 is forward gain.

S11 is SWR.

Calibrate an RF vector network analyzer with a short circuit, open circuit, and 50-ohm loads.

100W forward and 25W reflected is 75W absorbed by the load.

Q for a series tuned circuit can be determined by bandwidth.

Vector network analyzer can measure filter frequency response.

Measure IMD by modulating two non-harmonically related signals and observe on a spectrum analyzer.

Vector network analyzer can measure, input impedance, output impedance and reflection coefficient.

GROUP 4C – RECEIVER PERFORMANCE

Excessive phase noise can combine to generate interference.

Front-end filter or pre-selector to eliminate out-of-band signals.

Suppression in an FM receiver is capture effect.

Noise figure is ratio of receiver generated noise to theoretical minimum.

-174 dBm/Hz noise floor is perfect receiver at room temperature.

Increase bandwidth from 50 Hz to 1,000 Hz increases noise floor 13 dB.

MDS is minimum discernible signal.

SDR overloads when signals exceed reference voltage.

High frequency IF eliminates image responses.

Variety of IF bandwidths is good to match the mode.

Attenuator works because atmospheric noise is greater than internally generated noise.

Roofing filter attenuates signals near the receive frequency.

Reciprocal mixing is local oscillator phase noise mixing with strong signals.

IF shift is to reduce interference.

GROUP 4D – RECEIVER PERFORMANCE

Blocking dynamic range is gain compression.

Poor dynamic range causes desensitization.

Intermodulation interference is output signals mix in the final amplifier.

Reduce intermod with properly terminated circulator.

Intermod produced on average of two frequencies.

AMATEUR PRACTICES

To reduce desensitization, insert attenuation.

Intermod caused by nonlinear circuits or devices.

Preselector rejects signals outside of desired band.

Third-order intercept is when third-order intermod is same level as input signals.

Odd-order intermod is in the band of interest.

Link margin is the sum of power and path losses.

-100 dBm is 0.1 picowatts.

GROUP 4E – NOISE AND INTERFERENCE

Auto notch filtering removes CW and the interference.

Noise reduction works on broadband white noise, ignition noise and power line noise.

Noise blanker removes impulse noise.

Conducted noise reduced by ferrite chokes.

Motor noise reduced by brute-force AC-line filter.

Computer network equipment unstable signals at specific frequencies.

Shielded cable radiates from common-mode currents.

Current flowing equally on all conductors is common-mode.

Intermittent buzz on AC line is arcing contacts, doorbell transformer, and illuminated display.

Local AM on the HF bands caused by corroded metal.

Series of carriers at regular intervals is switch-mode power supply.

Install surge suppressor at single point ground panel.

Single point assures all lightning protectors activate at the same time.

E5 – ELECTRICAL PRINCIPLES

E5A Resonance and Q

The voltage across reactance in a series RLC circuit can be larger than the voltage applied due to *resonance*. Those high voltages can cause arcing in your antenna tuner.

The resonate frequency of a series RLC circuit if R is 22 ohms, L is 50 microhenrys and C is 40 picofarads is 3.56 MHz.

The resonate frequency of a parallel RLC circuit if the R is 33 ohms, L is 50 microhenrys and C is 10 picofarads is 7.12 MHz.

The formula for resonant frequency is $F=1/(2\pi \times \sqrt{LC})$. If that math is too much for you, as it is for me, there are online calculators, so I don't feel bad about giving you this cheat: *The resonant-frequency answer is either in the 80 or 40-meter ham band. All the other answers are elsewhere.*

The magnitude of the impedance in a series RLC circuit at resonance is approximately *equal to the circuit resistance.* *Hint: "RLC" refers to resistor, inductor, and capacitor. If the inductor and capacitor cancel each other out (resonance), all that is left is resistance.*

The magnitude of impedance with a resistor, inductor, and capacitor all in parallel at resonance is *equal to the circuit resistance.**
Hint: In series, or parallel, at resonance, the answer is the same. All that is left is the resistance.

"Q" stands for reactive quotient and is a measure of selectivity. A high Q circuit is very selective and narrow banded. **The effect of increasing the Q in an impedance matching network is the matching bandwidth is decreased.**

The magnitude of the circulating current in a parallel LC circuit at resonance is *maximum*. *Hint: The <u>circulating</u> current in the circuit is maximum. The inductive and capacitive reactance are in parallel, and the current gets transferred back and forth between them (circulating).*

The magnitude of the current at the input of a parallel RLC circuit at resonance is *minimum*. *Hint: The circulating current stays in the circuit and doesn't draw current from the supply. The glass is full. The <u>input</u> current is at a minimum.*

The phase relationship between the voltage and current in a series resonate circuit is the voltage and current are *in phase*. *Hint: In a resonant circuit, inductance and capacitance cancel each other, and voltage and current are in phase.*

The Q of an RLC parallel resonant circuit is calculated by *resistance divided by reactance*.

The half-power bandwidth of a resonant circuit that has a resonate frequency of 7.1 MHz, and a Q of 150 is 47.3 kHz. Half-power bandwidth is frequency divided by Q.
Solve: 7.1 MHz/150 = .0473 MHz = 47.3 kHz. The answers are far enough apart that you can get sloppy with your decimal points. Look for the solution with the correct integers.

The half-power bandwidth of a resonate circuit that has a resonant frequency of 3.7 MHz, and a Q of 118 is 31.4 kHz.
Solve: 3.7 MHz/118 = .3135 MHz is 31.35 kHz and 31.4 kHz is the closest.

Increasing Q in a series resonant circuit *increases internal voltages.* *Hint: Increasing Q in a series circuit increases reactance, which increases voltages. In a series circuit, Q=reactance/resistance.*

E5B Time Constants and Phase

The term for the time required for the capacitor in an RC circuit to be charged to 63.2% of the applied voltage or to discharge to 36,8% of its initial voltage is "*one time constant.*"

The time constant of a circuit having two 220-microfarad capacitors and two 1-megohm resistors in parallel is 220 seconds. : The formula is resistance times capacitance. *Solve: Two 220-microfarad capacitors in parallel equal 440 microfarads. Two 1-megohm resistors in parallel equal .5 megohms. 440 x .5 = 220.   Cheat: Remember, the answer is the value of the capacitor.*

The letter commonly used to represent *susceptance is Letter B.* Susceptance is the measure of how much a circuit is *susceptible* to conducting a changing current.

Impedance in polar form converted to admittance is the reciprocal of the magnitude and *change the sign of the angle.* *Hint: Admittance is the inverse or opposite. Take the reciprocal of the magnitude and change the sign of the angle.*

Admittance is the inverse of *impedance.* If the magnitude of a pure reactance is converted to susceptance it becomes the *reciprocal.* Reactance is a measure of how a circuit reacts to AC current and susceptance is its opposite. Susceptance is the reciprocal of reactance (1/X).

ELECTRICAL PRINCIPLES

Susceptance is the imaginary part of admittance.
Cheat: The only imaginary answer on the test.

Impedance makes the voltage or current lead one another. In an inductive circuit, voltage leads the current. In a capacitive circuit, voltage lags the current. Remember, "ELI the ICEman." E is voltage, L is inductance, I is current, C is capacitive. A capacitor "charges up" with the current leading. As the capacitor "fills" the current drops and voltage rises. The magnetic field around an inductor resists the flow of current and voltage leads.

The relationship between current through a capacitor and the voltage across it is: the current leads the voltage by 90 degrees. *Hint: Current leads voltage in a capacitor. ELI the ICEman told me so. Don't worry about the number of degrees.*

The relationship between the current through and the voltage across an inductor is voltage leads the current by 90 degrees. *Hint: Voltage leads current in an inductor says ELI.*

The phase angle between the voltage across and the current through a series RLC circuit if the XC is 300 ohms, the R is 100 ohms and XL is 100 ohms is 63 degrees with the voltage lagging the current. *Hint: The circuit is more capacitive, so the ICEman says voltage is lagging. We know the voltage is lagging the current because the XC is greater than the XL. That eliminates two answers. The rest of the math is staggering. Cheat: If XC is 300, the answer is 63.*

The phase angle between the voltage across and the current through a series RLC circuit if the XC is 25 ohms, the R is 100 ohms and XL is 75 ohms is 27 degrees with the voltage lagging the current. *Hint: We know the voltage is lagging the* *current because the XC is greater than the XL. That eliminates two answers. Cheat: If XC is 25, the answer is 27.*

E5C Coordinate Systems and Phasors

Impedances are described in polar coordinates by *magnitude and phase angle*.

The coordinate system used to display the phase angle of a circuit containing resistance, inductance and/or capacitive reactance is *polar coordinates*. *Memory point: Phase angle goes with polar coordinates.*

A purely inductive reactance in polar coordinates is a *positive 90 degree* phase angle. Positive because voltage is leading in an inductive circuit. Pure inductance will be 90 degrees.

The Y-axis scale most often used for graphs of circuit frequency response is *logarithmic*.

The diagram used to show the phase relationship between impedances at a given frequency is called a *phasor* diagram. *Cheat: If you see a Star Trek weapon in the answer, it is correct.*

A capacitive reactance of 100 ohms in rectangular notation is 0-j100. *Hint: Look at ELI the ICEman's voltage. This is a capacitive circuit, so the voltage is behind, -j.*

An impedance of 50-j25 represents 50 ohms resistance with 25 ohms capacitive reactance. The first number is ohms resistance. The second is reactance. Since j is negative, we know the voltage is lagging, and the circuit must be capacitive.

ELECTRICAL PRINCIPLES

When using rectangular coordinates to graph impedance , the *X axis is resistive* and Y axis is the reactive component. Figure 5.1 helps visualize this.

The impedance of a pure resistive circuit is plotted on the *horizontal* axis.

The following questions refer to figure E5-1 on the next page:
The point representing a 400-ohm resistor and a 38 picofarad capacitor at 14 MHz is point 4. *Solve: We know the answer lies on the +400-ohm line of the horizontal axis (resistance). We know the circuit is capacitive, so it will have a minus sign and be on the lower quadrant on the vertical axis. Point 4 it is, by default.*

The point representing a circuit has a 300-ohm resistor, and an 18 microhenry inductor at 3.5 MHz is point 3. *Solve: Inductive reactance is 2π times frequency times inductance. 2x3.14x3.5x18= 395. The circuit is inductive. Look up on the positive Y side, starting at the 300-ohm point on the X axis. Point 3 is the correct answer, where 300 and 395 cross.*

The point representing a circuit that has a 300-ohm resistor and a 19 picofarad capacitor at 21.200 MHz is point 1. *Solve: On a vertical line off the 300-ohm horizontal axis, and the circuit is capacitive, so the answer is lower (-Y) quadrant. Point 1 by default. Cheat: 300-ohm resistor is either point 1 or 3. If the circuit is inductive, Y is positive, point 3. If the circuit is capacitive, Y is negative, point 1.*

Figure E5-1

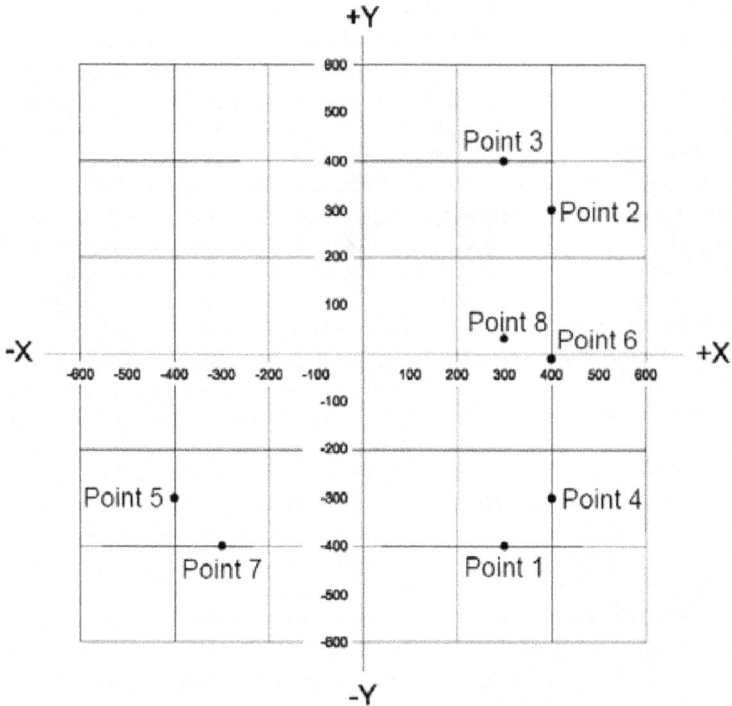

E5D RF Effects

The result of conductor skin effect is, *as frequency increases, resistance increases because RF current flows closer to the surface.* RF flows in a thinner layer of the conductor, closer to the surface.

It is important to keep lead lengths short for components in circuits for VHF and above *to minimize inductive reactance.*

The phase relationship for current and voltage for reactive power is *they are 90 degrees out of phase.*

ELECTRICAL PRINCIPLES

Short connections are used at microwave frequencies to *reduce phase shift along the connection.* *Hint: Another way to say "minimize reactance."*

A parasitic characteristic would be something about the design of the component that causes it to have an opposite and unintended effect. **The parasitic characteristic that causes electrolytic capacitors to be unsuitable for use at RF is *inductance.***

The parasitic characteristic that created an inductor's self-resonance is *inter-turn capacitance.* *Hint: the opposite effect.*

***A component's nominal and parasitic reactance* combine to create self-resonance.**

The primary cause of loss in film capacitors at RF is *skin effect.*

The reactive power in ideal inductors and capacitors is *stored in magnetic or electrical fields but not dissipated.* *Hint: If the components are "ideal," energy is not dissipated in them.*

Reactive power is *wattless, non-productive power.* It is the capacitor and inductor swapping power feeding each other.

As a conductor's diameter increases, *its electrical length increases.*

The real power consumed in a circuit consisting of a 100-ohm resistor in series with a 100-ohm inductive reactance drawing 1 ampere is *100 watts.* *This question assumes no loss in the inductor. How much power is consumed by the resistor portion of the circuit? The answer is $P=I^2R$. I^2 is 1; times the R (100) is 100 watts.*

SUMMARY: ELECTRICAL PRINCIPLES

GROUP 5A – RESONANCE AND Q

Voltage can be greater than applied due to resonance.

Resonant frequency answer is in the 40m or 80m band.

Resonant circuit impedance equal to circuit resistance.

Increasing Q decreases bandwidth.

Resonant circulating current in parallel is maximum.

Resonant input current in parallel is minimum.

Voltage and current are in phase in resonant circuit.

Q of parallel circuit is resistance divided by reactance.

Half-power bandwidth is frequency divided by Q.

Increasing Q in a series resonant circuit increases voltages.

GROUP 5B – TIME CONSTANTS AND PHASE

Time to charge to 63.2% is "one time constant."

Answer to time constant question is the same as capacitor value.

Susceptance is letter "B."

To convert impedance to admittance in polar form, reciprocal of the magnitude and change sign of angle.

Admittance is inverse of impedance.

Reactance converted to susceptance is the reciprocal.

Susceptance is imaginary part of admittance.

ELI the ICEMAN

GROUP 5C – COORDINATE SYSTEMS AND PHASORS

Impedance in polar coordinates is magnitude and phase angle.

Purely inductive reactance is positive 90 degree phase angle.

Y-axis graphs of frequency response are logarithmic.

Phase relationships and impedance shown on a phasor diagram.

ELECTRICAL PRINCIPLES

Capacitive reactance of 100 ohms is 0-j100.

50-j25 is 50 ohms resistance and 25 ohms capacitive.

X axis is resistive, Y axis is reactive.

Impedance of resistive plotted on horizontal axis.

GROUP 5D – RF EFFECTS

Skin effect means as frequency increases, resistance increases.

Short leads to minimize inductive reactance.

Phase relationship for reactive power is current and voltage are 90 degrees out of phase.

Short connections to reduce phase shift.

Electrolytic capacitors are unsuitable at RF because of inductance.

Nominal and parasitic reactance create self-resonance.

Loss in film capacitors due to skin effect.

Reactive power is stored in electrical or magnetic fields and not dissipated.

Reactive power is wattless, non-productive.

Electrical length increases as conductor diameter increases.

$P=I^2R$.

E6 – CIRCUIT COMPONENTS

E6A Semiconductors

Gallium arsenide is used as a semiconductor material in *microwave* circuits.

The semiconductor material that contains excess free electrons is *N-type*. *Hint: Excess free electrons would be **N**egatively charged.*

A PN-junction does not conduct current when reverse biased because holes in the P-type material and electrons in the N-type material are separated by the applied voltage *widening the depletion region*. *Hint: Reverse bias separates.*

The name given to an impurity that adds holes to a semiconductor crystal structure is *acceptor impurity*. *Hint: Holes accept electrons.*

The DC input impedance at the gate of a field-effect transistor is *higher* than a bipolar transistor. FETs have high input impedance.

The beta of a bipolar transistor is the *change in collector current with regard to the change in base current*. The change between the collector and base. *Cheat: Beta and base current.*

A silicon NPN transistor is biased on when the *base-to-emitter voltage is approximately .6 - .7 volts*. *Hint: It is biased by voltage, eliminating 2 of the answers. The bias voltage is small, eliminating the other answer.*

The term indicating the frequency at which a grounded-base current gain of a transistor has decreased to .7 of the gain obtainable at 1 kHz is

called the *alpha cutoff frequency.* Hint: Gain *(alpha) is being cut off (decreased).*

A depletion-mode FET is *an FET that exhibits a current flow between source and drain when no gate voltage is applied.* Hint: *Current flow with no gate voltage would deplete.*

There are the only two symbols you need to know from Figure E6-1. **In Figure E6-1, the symbol for an N-channel dual-gate MOSFET is *number 4.*** Hint: *It is dual gate and only two of the figures show two gates (G1 and G2). It is N channel so the arrow is pointed in.*

Figure E6-1

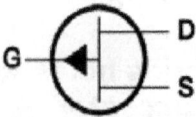

1

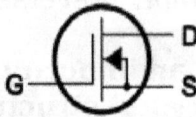

2

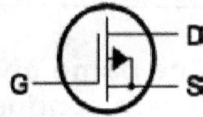

3

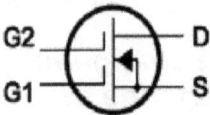

4

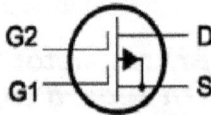

5

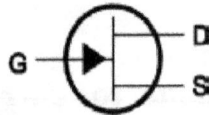

6

The symbol for a P-channel junction FET is *number 1.* Hint: *A P-channel FET has the arrow pointed out.*

The purpose of connecting Zener diodes between a MOSFET gate and its source or drain is *to protect the gate from static damage.*

E6B Diodes

The most useful characteristic of a Zener diode is *constant voltage drop under conditions of varying current.*

The important characteristic of a Schottky diode is *lower forward voltage drop.*
Hint: Constant and lower voltage drop is good.

The property of an LED semiconductor material that determines its forward voltage drop is *band gap.* Band gap is the minimum energy to excite an electron to conduct. When the LED conducts, current increases and Ohm's law tells us voltage drops.

The semiconductor device designed for use as a voltage-controlled capacitor is a *varactor diode.*
Hint: The voltage varies the capacitance.

The characteristic of a PIN diode that makes it useful as an RF switch is *low junction capacitance.* Hint: If it had high capacitance, the charge would hinder the switching ability.

 A common use of a Schottky diode is as a *VHF/UHF mixer or detector.*

A type of Schottky barrier diode is *metal semi-conductor junction.*

A common use for a point-contact diode is as an *RF detector.* Hint: An RF detector helps you make a contact.

To control the attenuation of RF signals by a PIN diode, use *forward DC bias current.* Hint: Give the signals a boost with forward bias.

CIRCUIT COMPONENTS

When a junction diode fails from excessive current, the cause is *excessive junction temperature.* Hint: *Excessive current causes excessive heat, which causes a failure.*

The schematic symbol for a Schottky diode in Figure E6-2 is *number 6.* This is the only symbol you need to remember from E6-2.

Figure E6-2

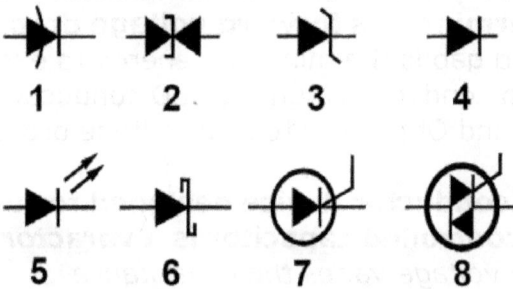

E6C Digital ICs

The function of hysteresis in a comparator is *to prevent input noise from causing unstable output signals.* Hysteresis is a voltage lag between active and inactive states. A 12-volt relay clicks on at 11 volts but doesn't click off until voltage drops to 9 volts. Small variations in voltage don't trigger it.

A comparator is a device that compares two voltages or currents and outputs a digital signal indicating which is larger. **When the level of a comparator's input signal crosses the threshold, *the comparator changes its output state.***

Tri-state logic is logic devices with *0,1 and high impedance output states.* Hint: *Tri-state is three output states: 0, 1 and high.*

BiCMOS logic is an integrated circuit logic family using both bipolar and CMOS transistors (Complementary Metal Oxide Semiconductor). **An advantage of BiCMOS logic is it has the *high input impedance of CMOS and the low output impedance of bipolar transistors.*** Hint: High input impedance is a good thing because it doesn't load down the circuit.

The digital logic family with the lowest power consumption is *CMOS* .

CMOS digital integrated circuits have high immunity to noise on the input signal or power supply because the *input switching threshold is about half the power supply voltage.* Hint: It switches long before power supply voltage variations have any effect.

A pull-up or pull-down resistor *is connected to the positive or negative supply line used to establish a voltage when an input or output is an open circuit.* Hint: Pull up or pull down means it could be positive or negative. When the circuit is open, the resistor sets the bias voltage.

The configuration of a field-programmable gate array (FPGA) is designed with *hardware descriptor language (HDL).* Hint: Design a gate with hardware.

Figure E6-3 has symbols for logic circuits. You only need to recognize numbers 2, 4 and 5.

In Figure E6-3, the schematic symbol for a NAND gate is number 2. Hint: Two inputs must be the same (AND) and the output is reversed (indicated by the little circle. Two high inputs will result in a low output.

CIRCUIT COMPONENTS

In Figure E6-3, the schematic symbol for a NOR gate is *number 4*. *Hint: The inputs are on a curved line (OR), and the output has the little circle indicating reversed output. It gives a positive output only if both inputs are negative .*

Figure E6-3

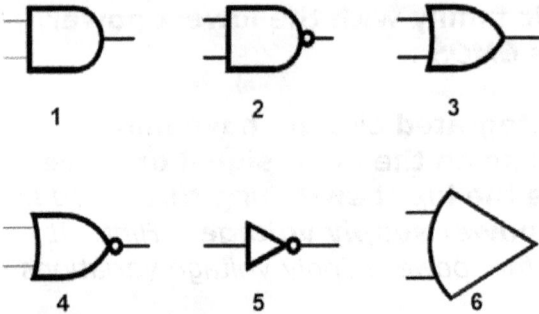

The symbol for the NOT operation (inversion) is *number 5.* *Hint: One input and one output are all that is required to invert (shown by the little circle).*

E6D Inductors and Piezoelectricity;

What is piezoelectricity? *A characteristic of materials that generate a voltage when stressed and that flex when a voltage is applied.*
Hint: When stressed; as a grill lighter sparks.

The equivalent circuit of a quartz crystal is *series RLC in parallel with a shunt C representing electrode and stray capacitance.* *Hint: The one answer without resonance. Resonance has nothing to do with piezoelectricity*

An aspect of the piezoelectric effect is *mechanical deformation of material by the application of a voltage.* *Hint: We usually think of it the other way around, a voltage generated by stressing the material.*

The cores of inductors and transformers are sometimes constructed of thin layers to *reduce power loss from eddy currents in the core.* Eddy currents are loops of electrical current caused by changing magnetic fields. Thin layers in the core reduce the loops.

Ferrite cores and powdered iron in an inductor compared is *ferrite toroids require fewer turns to produce a given inductance value.* Hint: Ferrite is more efficient.

The core material property that determines the inductance of an inductor is *permeability.* Permeability determines how the magnetic flux reacts. Think of it as a measure of the efficiency of the core.

The current in the primary winding of a transformer with no load attached is called *magnetizing current.* Hint: It is just enough to magnetize the core but no more because there is no load to draw current.

The material with the highest temperature stability of its magnetic characteristics is *powdered iron.* Hint: Iron can handle heat

The devices commonly used as VHF and UHF parasitic suppressors at the input and output terminals of a transistor HF amplifier are *ferrite beads.* Hint: Ferrite beads are toroids. The beads act as chokes.

The primary advantage of using a toroidal core over a solenoidal core is the toroidal core *confines most of the magnetic field within the core material.* A solenoidal core is a bar of ferrite with the wire wrapped around. The circular toroid keeps the magnetic field contained and concentrated.

CIRCUIT COMPONENTS

The material that decreases inductance when inserted into a coil is *brass*. *Hint: Brass has lower permeability and does not concentrate the magnetic lines as effectively as air.*

The primary cause of inductor saturation is *operating at excessive magnetic flux*. Magnetic flux is a measure of the magnetic field. Excessive flux saturates the core.

E6E Semiconductor Materials and Packages

Gallium arsenide (GaAs) is useful for semiconductor devices operating at UHF and higher frequencies because it has *higher electron mobility*. *Hint: Electron mobility is a strange enough term it should stick with you.*

An example of a through-hole device is a *DIP*. Dual Inline Package is a type of computer chip. The leads are soldered through holes in the circuit board.

MMIC means Monolithic Microwave Integrated Circuit. **The material likely to provide the highest frequency of operation when used in MMICS is *gallium nitride*.** *Hint: Gallium is used at high frequencies. Gallium is also the answer to other questions.*

The most common input and output impedance of a circuit that use MMICs is 50 ohms. *Hint: Just like coax.*

The noise figure typical for a low-noise UHF preamplifier is *0.5 dB*. *Hint: The pre-amplifier adds some noise, but not a lot. The other answers are negative or too high.*

The characteristic of MMIC that makes it popular for VHF through microwave circuits is *controlled gain, low noise figure, and constant input and output impedance over the specified frequency range*. Hint: You don't need to remember all that. Just pick "low noise."

The transmission line used for connections to MMICs is *microstrip*. Hint: Microwave = microstrip.

Power is supplied to an MMIC through a *resistor and/or RF choke connected to the amplifier output lead.* Hint: Power comes from the output lead just like power is applied to the plate of a tube.

The component package with the least parasitic effect above the HF range are *surface mount.* Hint: Surface mount has the shortest leads (none), and short leads reduce stray reactance at high frequencies.

The advantage of surface-mount technology over through-hole components is:
- *Smaller circuit area*
- *Shorter circuit-board traces*
- *Components have less parasitic inductance and capacitance*
- *All these choices are correct*

Hint: Recognize two, and you know the answer is all.

A characteristic of DIP packaging is *two rows of connecting pins placed on opposite sides of the package.* Hint: Dual Inline Package.

DIP through-hole package ICs are not typically used at UHF and higher frequencies because of *excessive lead length.* The leads are shorter than individual components might be but still longer than surface mount.

E6F Electro-optical Technology

Electrons absorb the energy from light falling on a photovoltaic cell.

When light shines on a photoconductive material the **resistance decreases**. *Hint: Photoconductive.*

The most common configuration of an optoisolator or optocoupler is an **LED and a phototransistor.** Two circuits are isolated (not electrically connected), but signals pass via light.

The photovoltaic effect is the **conversion of light to electrical energy.**

An optical shaft encoder **detects rotation by interrupting a light source with a patterned wheel.** *Hint: It senses light pulses when the shaft turns.*

A material most commonly used to create photoconductive devices is a **crystalline semiconductor.**

A solid-state relay is a **device that uses semiconductors to implement the functions of an electromechanical relay.** *Hint: A solid-state relay is a relay made with solid-state devices (semiconductors).*

An optoisolator is often used in conjunction with solid state circuits when switching 120VAC because optoisolators **provide an electrical isolation between the control circuit and the circuit being switched.** *Hint: Optoisolators provide isolation. Look for that word in the answer.*

The efficiency of a photovoltaic cell is the **relative fraction of light that is converted to**

CIRCUIT COMPONENTS

current. Hint: The more efficient, the more current it produces for the same amount of light.

The most common type of photovoltaic cell used for electrical power is *silicon*. Hint: Solar panels are made of silicon.

The approximate voltage produced by a fully illuminated silicon photovoltaic cell is *0.5 V*. Each cell produces 0.5 volts. Connect the cells in series to output a higher voltage.

SUMMARY: CIRCUIT COMPONENTS

GROUP 6A – SEMICONDUCTORS

Gallium arsenide is used in microwave circuits.

Excess electrons is N-type.

Reverse bias does not conduct because the applied voltage widens the depletion region;.

Impurity added is acceptor impurity.

Field-effect transistor has higher gate impedance than bi-polar.

Beta is change in collector current with regard to base.

Silicon NPN transistor biased at .6 - .7 volts.

Alpha cutoff frequency is .7 of gain at 1 kHz.

Depletion mode is current flow when no gate voltage.

N-channel dual gate has 2 gate and arrow pointed in.

Zener diodes in Mosfet to protect from static damage.

GROUP 6B – DIODES

Zener diode has constant voltage drop.

Schottky diode has lower forward voltage drop.

Varactor diode is a voltage-controlled capacitor.

PIN diode has low junction capacitance.

Schottky diode used in VHF/UHF mixer.

Metal semi-conductor junction is Schottky diode.

Point-contact diode is an RF detector.

CIRCUIT COMPONENTS

To control PIN diode attenuation, use forward bias current.

Diode fails from excessive junction temperature.

GROUP 6C – DIGITAL ICS

Hysterisis is to prevent unstable output signals.

Comparator changes output state when input level crosses threshold.

Tri-state logic is 0,1, and high.

BiCMOS has high input and low output impedance.

Lowest power consumption is CMOS.

CMOS input switching is half power supply voltage.

Pull-up or pull-down resistor connected to positive or negative supply.

Field-programmable gate array uses Hardware Descriptor Language (HDL).

NAND gate has 2 inputs on flat side and circle on output.

NOR gate has 2 inputs on curved side and circle on output.

NOT gate is one input and output.

GROOUP 6D – INDUCTORS AND PIEZOELECTRICITY

Piezoelectricity – material generates voltage when stressed

Equivalent circuit for a quartz crystal has nothing to do with resonance.

Piezoelectric effect is deformation by application of voltage.

Transformer cores in thin layers to reduce power loss from eddy currents.

Ferrite requires fewer turns than powdered iron.

Permeability of core material determines inductance.

No load current in primary is magnetizing current.

Powdered iron has highest temperature stability.

Parasitic suppressors are ferrite beads.

Toroid core confines the magnetic field within the core material.

Decrease inductance by inserting brass.

Inductor saturation caused by excessive magnetic flux.

GROUP 6E – SEMICONDUCTOR MATERIALS

Gallium arsenide has higher electron mobility.

Through-hole device is a DIP.

Galium nitride for highest frequency of operation.

Common MMIC impedance is 50 ohms.

Noise figure for UHF pre-amp is 0.5 dB.

MMIC is popular for its low noise.

MMIC transmission line is microstrip.

Power is supplied through the amplifier output lead.

Lest parasitic effect is surface mount.

Surface mount advantages: small area, shorter board traces, and less parasitic inductance and capacitance.

DIP packaging is two rows of pins.

DIP not used at UHF because of excessive lead length.

GROUP 6F – ELECTRO-OPTICAL TECHNOLOGY

Electrons absorb energy from light on photovoltaic cell.

Light shining on photoconductive decreases resistance.

Optoisolator is LED and phototransistor.

Photovoltaic is conversion of light to electrical.

Optical shaft encoder interrupts light source with patterned wheel.

Photoconductive devices use crystalline semiconductor.

Solid-state relay uses semiconductors instead of electromechanical relay.

Optoisolator provides electrical isolation.

Efficiency of photocell is relative fraction of light converted.

Common material is silicon.

Cell is 0.5 volts.

E7 – PRACTICAL CIRCUITS

E7A Digital Circuits

A bi-stable circuit is a *flip-flop*. *Hint: "Bi" means it has two states, flip and flop.*

The function of a decade counter digital IC is to *produce one output signal for every ten input pulses*. *Hint: It counts decades or tens.*

The circuit that can divide the frequency of a pulse train by 2 is a *flip-flop*.

To divide a signal frequency by 16 requires 4 flip-flops. *Hint: 2 X2 X2 X2.*

A circuit that continuously alternates between two states without an external clock is an *astable multivibrator*. *Hint: Astable means it alternates continuously. It is never stable.*

A monostable multivibrator *switches temporarily to an alternate state for a set time*. *Hint: Monostable means it switched temporarily and returns to one state.*

A NAND gate *produces a 0 at its output only if all inputs are 1*. *Hint: N means it produces an opposite logic. AND means both inputs are the same.*

An OR gate *produces a 1 if any or all inputs are 1*. *Hint: There is no N in front, so output equals input. OR means either input could be 1.*

A two-input exclusive NOR gate *produces a logic 0 if one and only one input is logic 1*. *Hint: The N means it produces an opposite logic. The OR means either input could be 1.*

A truth table is a *list of inputs and corresponding outputs for a digital device.* Hint: *It is a table showing the various outcomes for a digital device.*

"Positive logic" means *high voltage represents a 1, low voltage 0.* Hint: *1 is a positive number.*

E7B Amplifiers

Amplifiers can operate over all or part of a 360-degree signal cycle. Their Class describes how much of the cycle conducts.

A push-pull Class AB amplifier conducts *more than 180 degrees but less than 360 degrees.*

A Class D amplifier *uses switching technology to achieve high efficiency.*

The circuit required at the output of an RF switching amplifier is *a filter to remove harmonic content.* Hint: *Switching circuits can generate harmonics and the filter removes them.*

The operating point of a Class A common emitter amplifier is *halfway between saturation and cutoff.* Hint: *Class A runs throughout the 360-degree cycle, so it would have to be biased right in the middle.*

To prevent unwanted oscillations in an RF power amplifier, *install parasite suppressors and/or neutralize the stage.* Hint: *Suppressors suppress unwanted oscillations. Neutralization is introducing some negative feedback to cancel the oscillations.*

A characteristic of a grounded-grid amplifier is low input impedance. Hint: *If the grid is grounded, it is at low impedance.*

PRACTICAL CIRCUITS

The result of using a Class C amplifier to amplify single-sideband phone would be *signal distortion and excessive bandwidth.* Class C operates less than 50% of the time. Class C is not linear and is suited for CW but not SSB.

Switching amplifiers are more efficient than linear amplifiers because the *switching device is at saturation or cutoff most of the time.* *Hint: Another way of saying it only conducts for a short part of the cycle.*

Figure E7-1

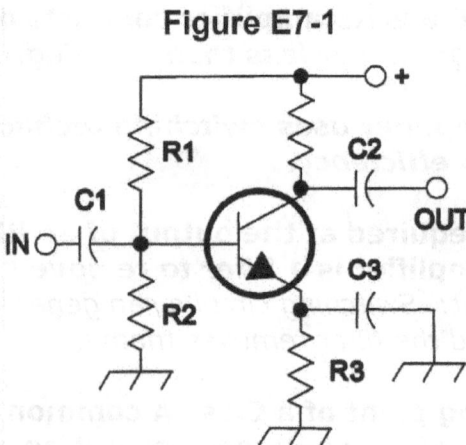

Figure E7-1 is a *common emitter transistor amplifier.*

There are three questions related to figure E7-1 and you only need to identify it as "common emitter" and know the purpose of R1/R2 and R3. **In Figure E7-1 the purpose of R1 and R2 is** *voltage divider bias.* They form a voltage divider between the + and ground to set the bias.

The purpose of R3 is *self bias.* Amplifiers operate with varying signal inputs. Biasing establishes the correct operating point of the transistor and, if done correctly, reduces distortion.

E7C Filters and Matching Networks

Capacitors and inductors in a low-pass filter Pi-network are arranged *with a capacitor connected between the input and ground, another capacitor between the output and ground and an inductor is connected between the input and output.* Hint: Another overly-complicated answer. The filter is to pass low frequencies, so it must shunt off high frequencies. The way to do that is with capacitors to ground from both the input and output. Look for that as part of the answer.

The frequency response of a T-network with series capacitors and a parallel shunt inductor is high-pass. Hint: Series capacitors pass high frequencies so it is a high-pass filter. The opposite of the above.

The purpose of adding an inductor to a Pi-network to create a Pi-L-network is *greater harmonic suppression.* Hint: The extra component (inductor) offers more suppression.

An impedance matching circuit transforms a complex impedance to a resistive impedance by *canceling the reactive part of the impedance and changing the resistive part to a desired value.* Hint: Impedance matching cancels reactance and changes to a desired value. Look for "a desired value" in the answer.

The filter with a ripple in the passband and a sharp cutoff is a *Chebyshev filter.* Cheat: Recognize the Russian name. It is only in one answer.

The distinguishing features of an elliptical filter are *extremely sharp cutoff with one or more notches in the stop band.* Cheat: "Extremely sharp cutoff" is good in a filter and the answer.

PRACTICAL CIRCUITS

A **Pi-L network** is a *Pi network with an additional output series inductor.* *Hint: Pi-L adds an L (inductor) to a Pi network.*

The filter most frequently used as a band-pass or notch filter on VHF and UHF transceivers is a **helical filter.**

A **crystal lattice filter** is a *filter for low level signals made using quartz crystals.* Several crystals are connected in a lattice making the filter narrower and sharper. *Hint: The crystals aren't lattice-shaped; the circuit is. It is for a receiver, hence low-level signals.*

The filter for a 2-meter repeater duplexer would be a *cavity filter.* Cavity filters are sharply tuned resonant circuits that allow only the design frequency to pass. A duplexer isolates the receiver from the transmitter when they both use the same antenna.

The term used to describe a filter's ability to reject adjacent signals is *shape factor.* A narrow shape factor has steeper sides.

E7D Power Supplies

A **linear electronic voltage regulator** works by the *conduction of a control element varied to maintain a constant output voltage.* *Hint: It is a regulator, so it regulates to "maintain a constant output voltage."*

A **switching voltage regulator** works by *varying the duty cycle of pulses input into a filter.* *Hint: Switching is changing the duty cycle to produce a constant average.*

The device typically used as a stable reference voltage in a linear voltage regulator is a *Zener diode.* Zener diodes are voltage regulators.

A three-terminal voltage regulator is a *series regulator.* *Hint: one terminal in, one out and one to ground placed in series between the supply and load.*

The linear voltage regulator that operates by loading the unregulated voltage source is a *shunt regulator.* *Hint: The shunt keeps a constant load by dumping excess to ground.*

There are three possible questions related to Figure E7-2. **The circuit shown in Figure E7-2 is a *linear voltage regulator.***

Figure E7-2

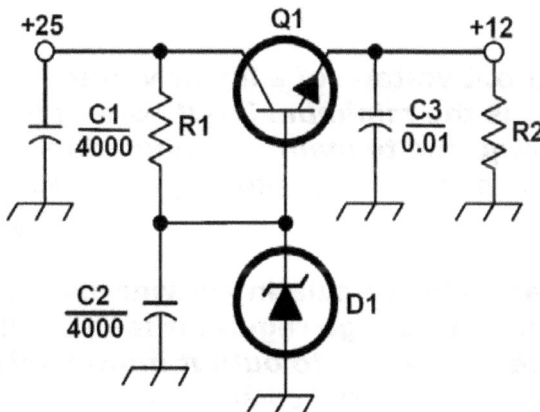

The purpose of Q1 in Figure E7-2 is to *control the current to keep the output voltage constant.* *Hint: The current is passing through Q1, a big power transistor, instead of the regulator. Q1 is between the input and output.*

PRACTICAL CIRCUITS

The purpose of C2 in Figure E7-2 is to *bypass rectifier output ripple around D1.* Hint: *Capacitors are often used to bypass ripple (hum).*

Battery time is calculated by *capacity in amp-hours divided by average current.* A 20 Ah battery should supply 2 amps for 10 hours.

A switching type high voltage power supply is less expensive and lighter than an equivalent linear power supply because *the high-frequency inverter design uses much smaller transformers and filter components for an equivalent power output.* You can get the same inductance and capacitance in smaller components at higher frequencies.

The purpose of an inverter connected to a solar panel is to *convert the panel's output from DC to AC.*

The drop-out voltage of a linear voltage regulator is the *minimum input-to-output voltage required to maintain regulation.* Hint: *Without the minimum input-to-output voltage, it drops out and stops regulating.*

The equation for calculating power dissipation by a series linear voltage regulator is the *voltage difference from input to output multiplied by output current.* Hint: *You are solving for dissipation, the power kept in the regulator. That depends on the difference is between input and output. P=IE.*

The purpose of connecting equal-value resistors across power supply filter capacitors connected in series is to:
  • *Equalize the voltage across each capacitor*

- *Discharge the capacitors when voltage is removed*
- *Provide a minimum load to the supply*
- *All these choices are correct*

The purpose of a "step-start" circuit in a high-voltage power supply is *to allow the filter capacitors to charge gradually.* Hint: *It starts in steps to avoid a possibly damaging voltage surge.*

E7E Modulation and Demodulation

FM phone signals can be generated by *reactance modulation of a local oscillator.* Hint: *It is FM, so you modulate the oscillator, and a reactance modulator is one way to do that.*

The function of a reactance modulator is to *produce PM or FM signals by varying a capacitor.* Hint: *"Reactance modulator and FM." Vary the phase or frequency with reactance from capacitors.*

A frequency discriminator stage in an FM receiver is a *circuit for detecting FM signals.*

One way to generate a single-sideband phone signal is using a *balanced modulator followed by a filter.* Hint: *You filter off the other sideband.*

The circuit added to an FM speech channel to boost the higher audio frequencies is a *pre-emphasis network.* Hint: *It emphasizes the higher audio frequencies.*

De-emphasis is used in FM communications receivers *for compatibility with transmitters using phase modulation.* It undoes the emphasis added by a pre-emphasis circuit and makes the signal compatible.

PRACTICAL CIRCUITS

The term "baseband" means *the frequency range occupied by the message signal prior to modulation.* Hint: The base frequency range before modulation.

The principal frequencies that appear at the output of a mixer are the *two input frequencies along with their sum and difference frequencies.* Hint: It is a mixer circuit, so there are two input frequencies, and the result of the mixing is a sum and a difference.

When signal levels to a mixer are too high, *spurious mixer products* are generated. Hint: Overloading a circuit produces spurious products.

Rectification, detection, and demodulation are used interchangeably. The three terms have the same result when referring to receiver functions. A diode envelope detector functions by *rectification and filtering of RF signals.* Hint: A detector detects RF.

The detector used for demodulating SSB signals is called a *product detector*.

E7F Software Defined Radio and DSP

"Direct sampling" in software defined radios means *incoming RF is digitized* by an analog-to-digital converter without being mixed with a local oscillator signal. Hint: A software defined radio digitizes. It is direct because the RF is not first converted to an IF with a local oscillator, but you don't need to know that to answer the question.

The digital signal processing audio filter used to remove unwanted noise from a received SSB signal is an *adaptive filter.* Hint: Digital filters use *computing power to look for patterns and adapt to eliminate the noise.*

The digital processing filter used to generate an SSB signal is a *Hilbert-transform filter*. *Hint: Digital processing is transforming digits to audio. You need a transform filter.*

A method of generating an SSB signal using digital signal processing is *signals are combined in a quadrature phase relationship*.

To be accurately reproduced, an analog signal must be sampled by an analog-to-digital converter *at least twice the rate of the highest frequency component of the signal*. To convert from analog to digital requires double sampling.

The minimum number of bits required for an analog-to-digital converter to sample with a range of 1 volt at a resolution of 1 millivolt is *10 bits*. *Solve: The resolution is 1/1000 of the range. It takes 10 bits to count to a thousand. $2^{10} = 1024$.*

The function of a Fast Fourier Transform is to *convert digital signals from the time domain to the frequency domain*. *Cheat: It is fast, converting slow time to frequency.*

The function of decimation is *reducing the effective sample rate by removing samples*. When the Romans conquered, they would often kill every tenth enemy captured, a process called decimation. Decimation reduced the enemy's effectiveness by removing a sample. This has nothing to do with radio, but it might help you remember.

An anti-aliasing digital filter is required in a digital decimator *to remove high-frequency signal components which would otherwise be*

reproduced as lower frequency components.
Hint: Those lower-frequency components would be aliases of the signal.

In a direct-sampling software defined receiver, using analog-to-digital conversion, the maximum bandwidth is determined by *the sample rate.*
Hint: A higher sample rate allows for more conversions, which is more bandwidth.

The minimum detectable signal level for an SDR in the absence of atmospheric or thermal noise is set by the *reference voltage level and sample width in bits.* Cheat: Simplify, the signal level is set by voltage.

Finite Impulse Response (FIR) filters *delay all frequency components of the signal by the same amount.* Hint: The delay is "finite," a set amount that stays the same.

The function of taps in a digital signal processing filter is to *provide incremental signal delays for filter algorithms.* Hint: think of "taps" as time-out delays that allow the filter to catch up.

To create a sharper filter response in a digital processing filter, *use more taps.* Hint: More time to process.

E7G– Operational Amplifiers

The typical output impedance of an integrated circuit op-amp is *very low.* Hint: Low impedance means lots of current, which means lots of power, and that is what you want in an amplifier.

The typical input impedance of an integrated circuit op-amp is *very high.* Hint: High input impedance means it does not load down the circuit and only draws a little power on the input. Op-amps are high impedance input and low impedance output.

An operational amplifier is a *high-gain, direct-coupled differential amplifier with a very high input impedance and a very low output impedance.* Hint: High input and low output impedance.

The "op-amp offset voltage" is the *differential line voltage needed to bring the open loop output voltage to zero.* Hint: You adjust the offset to bring the device to zero.

To prevent ringing and audio instability in an op-amp audio filter circuit, *restrict both gain and Q.* Hint: Too too much gain will cause instability, and too narrow a filter (high-Q) will ring.

The gain-bandwidth of an operational amplifier is the *frequency at which the open-loop gain of the amplifier equals one.*

Figure E7-3 on the following page is an operational amplifier (OP-AMP)

If a capacitor is added across the feedback resistor in Figure E7-3, the frequency response is a *low-pass filter.* The capacitor has a higher reactance to low frequencies, increasing the resistance across RF and increasing the gain for the low frequencies.

In Figure E7-3, if R1 is 1,000 ohms, RF is 10,000 ohms and 0.23 volts DC is applied to the input, the output voltage would be *-2.3 volts.* Hint: RF is above R1 on the schematic and in the formula. The ratio of the resistors RF/R1 is 10, so the gain is 10 times. The output voltage of an op-amp is inverted. Hence the answer is -10 x .23 = -2.3 volts.

PRACTICAL CIRCUITS

The absolute voltage gain expected when R1 is 1,800 ohms and RF is 68 kilohms is 68,000 / 1,800 = 38. *Solve: It is the ratio of the two resistors. RF/R1 = 68,000 / 1,800 = 38.*

If R1 is 10 ohms and RF is 470 ohms, the gain is 47. The questions are about voltage gain, not voltage. There is no inverting, and luckily, there are no answers with negative numbers to distract you.

The absolute voltage gain expected from the circuit in E7-3 when R1 is 3300 ohms and RF is 47 kilohms ohms is 14. *RF/R1 = 47,000/3,300 = about 14.*

Figure E7-3

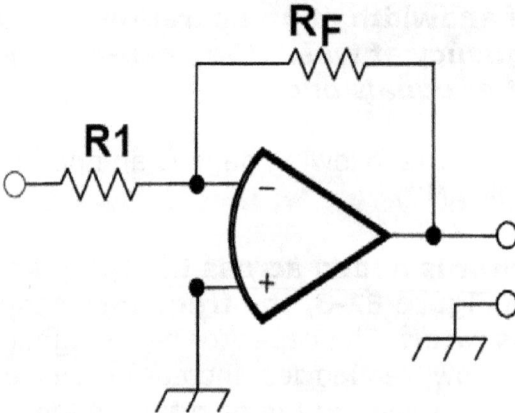

Figure E7-3 is an op-amp.

E7H Oscillators and Signal Sources

Three common oscillator circuits are *Colpitts, Hartley, and Pierce.* *Cheat: Colpitts is only in the correct answer.*

Positive feedback in a Hartley oscillator is supplied through a *tapped coil.* *Cheat: Har**T**ley and **T**apped coil.*

Positive feedback in a Colpitts oscillator is supplied through a *capacitive divider.* *Cheat: **C**olpitts and **C**apacitive.*

Positive feedback in a Pierce oscillator is supplied through a *quartz crystal.* *Cheat: Mind your Ps and Qs.*

A phase-locked loop is *an electronic servo loop consisting of a phase detector, a low-pass filter, a voltage-controlled oscillator and a* stable reference oscillator. *Hint: Phase-locked loops are stable. Look for the answer with "stable reference oscillator" and ditch the rest of this bloated answer.*

The functions that can be performed by a phase-locked loop are *frequency synthesis and FM demodulation.* *Hint: A loop is an oscillator, and only one answer has frequency synthesis.*

A microphonic is a *change in oscillator frequency due to mechanical vibration.*

Reduce an oscillator's microphonic responses by *mechanically isolating the oscillator from its enclosure.* *Hint: Keep it from mechanical vibration by mechanically isolating it.*

Components to reduce thermal drift in a crystal oscillator are *NP0 capacitors.* NP0 capacitors are

PRACTICAL CIRCUITS

made from ceramic and are very stable, so they don't change value when heated.

The frequency synthesizer that uses a phase accumulator, lookup table, digital to analog converter and a low-pass anti-alias filter is a *direct digital synthesizer.* *Hint: It has a digital to analog converter, so it must be a digital synthesizer.*

The information contained in the lookup table of a direct digital synthesizer (DDS) is the *amplitude values that represent the desired waveform.* *Hint: A table would contain "values." Look for the answer with "values."*

The major spectral impurity components of direct digital synthesizers are *spurious signals at discrete frequencies.* *Hint: Spectral impurity is caused by spurious signals.*

To insure that a crystal oscillator provides the frequency specified by the manufacturer, *provide the crystal with a specified parallel capacitance.* A crystal is designed for a particular frequency, but it can be "pulled" by stray capacitance. The manufacturer calibrates for a specified parallel capacitance.

A technique for providing highly accurate and stable oscillators for microwave transmission and reception is:
- *Use a GPS signal reference.*
- *Use a rubidium stabilized reference oscillator.*
- *Use a temperature-controlled high Q dielectric resonator.*
- *All these choices are correct.*

Hint: GPS and rubidium are easy to recognize, and when you have confidence in 2 out of 3, you can be safe selecting "all of the above."

SUMMARY: PRACTICAL CIRCUITS

GROUP 7A – DIGITAL CIRCUITS
Bi-stable circuit is a flip-flop.
Decade counter is 1 out for 10 in.
Divide frequency by 2 with a flip-flop.
To divide by 16 requires 4 flip-flops.
Continuous alternate between two states is astable multivibrator.
Monostable multivibrator switches temporarily for a set time.
NAND gate is 0 at the output if all inputs are 1.
OR gate is 1 if any or all inputs are 1.
Two-input NOR gate is 0 if one and only one input is 1.
Truth table is a list of inputs and outputs.
Positive logic means high voltage is a 1.

GROUP 7B -AMPLIFIERS
Push -pull more than 180 degrees.
Class D uses switching technology.
Requires filter to remove harmonics.
Class A operates halfway between saturation and cutoff.
To prevent unwanted oscillations, install suppressors and/or neutralize.
Grounded grid amplifier is low input impedance.
Class C would distort SSB.
Switching amplifiers more efficient because switching device is at saturation or cutoff most of the time.
Common emitter transistor amplifier.
Two tapped resistors for voltage divider bias.
Single resistor for self bias.

GROUP 7C – FILTERS AND MATCHING
Inductor between in and out is low pass.
Series capacitors is high pass.
Add inductor to Pi-network for greater suppression
Impedance matching circuit cancels reactive part.
Ripple in the passband is a Chebyshev filter.
Elliptical filter has sharp cutoff and notches.
Pi-L network is Pi with added inductor.
Band-pass or notch filter is helical.

PRACTICAL CIRCUITS

Crystal lattice filter for low-level signals.
2-meter duplexer uses cavity filter.
Filter's ability to reject adjacent signals is shape factor.

GROUP 7D – POWER SUPPLIES

Linear voltage regulator maintains a constant output voltage.
Switching voltage regulator varies duty cycle of pulses.
Stable reference voltage regulator uses Zener diode.
Three-terminal regulator is series regulator.
Voltage regulator loading the source is shunt regulator.
Circuit is a linear voltage regulator.
Q1 controls the current.
Capacitor bypasses ripple.
Battery time is amp-hours divided by average current.
Switching power supply uses smaller transformers and components.
Inverter on solar panel is to convert DC to AC.
Drop-out voltage is minimum required to maintain regulation.
Dissipation is difference in output and input voltage multiplied by output current.   P=IE.
Equal value resistors across filter capacitors are to equalize voltage, discharge when off, and provide minimum load.
Step-start circuit allows capacitors to charge gradually.

GROUP E7E- MODULATION AND DEMODULATION

FM generated by reactance modulation of local oscillator.
Reactance modulator varies a capacitor to produce FM.
Frequency discriminator is to detect FM signals.
Single-sideband generated by balanced modulator and filter.
Boost higher audio frequencies with pre-emphasis.
De-emphasis added on the receive side for compatibility.
Baseband means frequency range prior to modulation.

PRACTICAL CIRCUITS

Output of a mixer is two original frequencies plus sum and difference.

If signal levels to mixer are too high, it can produce spurious mixer products.

Diode envelope detector rectifies and filters.

Demodulate SSB with a product detector.

GROUP 7F – SDR AND DSP

Direct sampling does not mix with local oscillator.

DSP removes noise with adaptive filter.

SSB generated with Hilbert-transform filter.

SSB digital processing in quadrature phase relationship.

Analog to digital converter should sample two times the highest frequency.

Minimum bits at 1 volt with 1 millivolt resolution is 10 bits. $2^{10} = 1024$

Fast Fourier Transform from time domain to frequency domain.

Decimation removes samples.

Anti-aliasing removes high-frequency signal components.

SDR maximum bandwidth determined by sample rate.

Minimum detectable signal is set by voltage level.

Finite Impulse Response filters delay all components of the signal by the same amount.

Function of taps is to provide signal delays for filter algorithms.

To create a sharper response, use more taps.

GROUP 7G – OPERATIONAL AMPLIFIERS

Output impedance of op-amp is very low.

Input impedance of op-amp is very high.

Op-amp is very high gain.

Op-amp offset voltage brings the output voltage to zero.

To prevent ringing and instability, restrict gain and Q.

Gain-bandwidth is open-loop gain of one.

Capacitor added to op-amp is low-pass filter.

Amplification = top resistor divided by bottom.

Output voltage is inverted.

PRACTICAL CIRCUITS

GROUP E7H – OSCILLATORS AND SIGNAL SOURCES

Common oscillator is Colpitts.

Hartley uses feedback through a tapped coil.

Colpitts uses a capacitive divider.

Pierce uses a quartz crystal.

Phase-locked loops use a stable reference oscillator.

Phase-locked loop can do frequency synthesis and FM demodulation.

Microphonics is change in frequency due to mechanical vibration.

To reduce microphonics, isolate oscillator from enclosure.

Reduce drift with NPO capacitors.

Direct digital frequency synthesizer has a digital to analog converter.

Lookup table is amplitude values that represent the waveform

Major impurities are spurious signals.

Crystal filters specify a parallel capacitance.

Highly accurate oscillators use GPS reference, rubidium stabilized, and temperature controlled high-Q resonator.

E8 – SIGNALS AND EMISSIONS

E8A Waveforms

The process that shows a square wave is made up of a sine wave plus all its odd harmonics is _Fourier analysis._ Fourier analysis runs the signal through mathematical filters to extract its parts. A square wave contains odd harmonics.

A type of analog-to-digital conversion is _successive approximation._ _Hint: The converter goes through successive stages to calculate the end result._

A signal in the time domain is _amplitude at different times._

"Dither" in an analog-to-digital converter is a small amount of noise added to the input signal to _reduce quantization noise._ _Hint: Quantization._ Adding the noise prevents the converter from accumulating a quantity of errors over time.

The benefit of making voltage measurements with a true-RMS calculating meter is _RMS is measured for both sinusoidal and non-sinusoidal signals._ _Hint: Use a true meter to measure all signals._

The approximate ratio of PEP-to-average power in a typical single-sideband phone signal is _2.5 to 1._ _Hint Peak is higher than average but not crazy higher._

The factor that determines the PEP-to-average ratio of a single-sideband signal is the _speech_

characteristics. *Hint: Continuous and loud will put out more PEP power.*

A direct or flash conversion analog-to-digital converter is used for a software-defined radio because *very high speed allows digitizing high frequencies.* *Hint: "direct or flash" implies high speed.*

An analog-to-digital converter with an 8 bit resolution can encode *256* levels. *Solve 2^8 = 256.*

The purpose of a low-pass filter used at the output of a digital-to-analog converter is to *remove spurious sampling artifacts from the output signal.* *Hint: A low-pass filter removes harmonics, "spurious artifacts."*

A measure of the quality of an analog-to-digital converter is *total harmonic distortion.* *Hint: Low distortion would be quality.*

E8B Modulation and Demodulation

The modulation index of an FM signal is the ratio of *frequency deviation to modulating signal frequency.* *Hint: A "ratio" is sometimes called an "index." The correct answer compares two frequencies.*

The modulation index of a phase-modulated emission *does not depend* on the RF carrier frequency. *Hint: If it is phase modulated, the carrier frequency doesn't change*

The modulation index of an FM-phone signal having a maximum frequency deviation of 3000 Hz on either side of the carrier frequency when the modulating frequency is 1000 Hz is *3*. *Solve:*

The ratio of deviation (3000) and modulating frequency (1000) is 3.

The modulation index of an FM-phone signal having a maximum carrier deviation of plus or minus 6 kHz when the modulated with a 2 kHz modulating frequency is 3. *Solve: 6/2 = 3.*

Deviation ratio is the ratio of the *maximum carrier frequency deviation* (swing) *to the highest audio modulating frequency.*

The deviation ratio of an FM-phone signal having a maximum frequency swing of plus-or-minus 5 kHz when the maximum modulating frequency is 3 kHz is 5000/3000 = *1.67*

The deviation ratio of an FM-phone signal having a maximum frequency swing of plus or minus 7.5 kHz when the maximum modulation frequency is 3.5 kHz is 7.5/3.5 = *2.14.* *Hint: Calculate deviation ratio in the same manner as the modulation index: divide the larger number by the smaller.*

Orthogonal frequency division multiplexing (OFDM) is a technique used for *digital modes.* Data is split among several closely-spaced frequencies instead of a single wide-channel frequency. *Hint: Digital modes pass data.*

Orthogonal frequency division multiplexing (OFDM) is a digital modulation technique using *subcarriers at frequencies chosen to avoid intersymbol interference.* *Hint: "multiplexing" means it uses multiple subcarriers. Choose the answer with "subcarriers."*

Frequency division multiplexing (FDM) is *dividing the transmitted signal in separate frequency bands that each carry a different data*

stream. *Hint: Multiplexing is two or more. Look for "separate frequency bands."*

Digital time division multiplexing is two or more signals *arranged to share discrete time slots* of a data transmission. *Hint: Multiplexing "shares." Look for "time division" and "time slots."*

E8C Digital Signals

Quadrature Amplitude Modulation (QAM) is *transmission of data by modulating the amplitude of two carriers of the same frequency but 90 degrees out of phase.* *Hint: Quadrature is 90 degrees out of phase. Ditch the rest.*

The symbol rate in a digital transmission is the *rate at which the waveform changes to convey information.* *Hint: Rate of change*

Phase-shifting of a PSK signal should be done at the zero crossing of the RF signal to *minimize bandwidth.* *Hint: "Minimize bandwidth" is always good.*

The technique that minimizes the bandwidth of a PSK31 signal is *use of sinusoidal data pulses.* *Hint: Sinusoidal means a smooth and repetitive oscillation. "Smooth" sounds like it would minimize bandwidth. Look for the word "sinusoidal."*

The approximate bandwidth of a 13-WPM international Morse code transmission is *52 Hz.*

The bandwidth of an FT8 signal is *50 Hz.*

The necessary bandwidth of a 4800-Hz frequency shift 9600 baud ASCII FM transmission is *15.36 kHz.* *Hint: The widest answer.*

The way ARQ accomplishes error correction is *if errors are detected, retransmission is requested.* Hint: ARQ stands for Automatic Repeat reQuest.

The digital code which allows only one bit to change between sequential code values is *Gray code.* Hint: If you see "Gray code" in an answer, it is correct.

Data rate may be increased without increasing bandwidth by using a more efficient digital code. Hint: More efficient increases the rate.

The relationship between symbol rate and baud is that they are the same. Hint: The two words mean the same thing.

The factors that affect the bandwidth of a transmitted CW signal are *keying speed and shape factor* (rise and fall time). Hint: Faster digital modes are wider, so is faster CW.

The constellation diagram of a QAM or QPSK signal describes *the possible phase and amplitude states for each symbol.* Hint: It's a chart.

Nodes in a mesh network have *Internet Protocol (IP) addresses.* Mesh networks use repurposed computer wireless routers and they have IP addresses.

The technique individual nodes use to form a mesh network is *discovery and link establishment protocols.* Hint; They discover each other and link.

SIGNALS AND EMISSIONS

E8D Keying Defects and Overmodulation

Spread spectrum signals are resistant to interference because *signals not using the spread spectrum algorithm are suppressed in the receiver.* Hint: If a signal doesn't behave as expected, the receiver won't follow it.

The spread spectrum communications technique that uses a high-speed binary bit stream to shift the phase of an RF carrier is called *direct sequence.* Instead of changing frequency, a binary bit stream of extra data is injected directly into the signal. *Hint: Direct sequence stays on frequency.*

Spread spectrum frequency hopping *rapidly varies the frequency of a transmitted signal according to a pseudorandom sequence.* Hint: SS hops around with the receiver following the transmitter.

The primary effect of extremely short rise and fall times on a CW signal is the *generation of key clicks.*

The most common method of reducing key clicks is to *increase the keying waveform rise and fall times.*

The advantage of using a parity bit with an ASCII stream is *some types of errors can be detected.* The parity bit tells the receiver if the ASCII stream had an even or odd number of data bits. It is the simplest form of error detection.

A common cause of overmodulating AFSK signals is *excessive transmit audio levels.* Hint: AFSK is audio frequency-shift keying. If you overdrive the audio, you will overmodulate the signal.

The parameter that evaluates distortion of an AFSK signal caused by excessive input audio levels is *Intermodulation Distortion (IMD)*. *Hint: "Distortion" is in both the question and answer.*

An acceptable maximum IMD level for an idling PSK signal is -30dB. *Hint: An acceptable level has got to be negative. Only one answer is.*

Some of the differences between Baudot digital code and ASCII are *Baudot uses 5 data bits* per character, and ASCII uses 7 or 8. *Hint: The answer is longer, but all you need to recognize is that Baudot uses 5 data bits.*

An advantage to using ASCII for data transmissions is it is *possible to transmit both lowercase and uppercase*. Baudot is all caps with no lowercase letters.

SUMMARY: SIGNALS AND EMISSIONS

GROUP 8A – WAVEFORMS

Fourier analysis shows sa quare wave is sine and odd harmonics.

Analog to digital conversion is successive approximation.

Time domain is amplitude at different times.

Dither is to reduce quantization noise.

True-RMS voltmeter measures both sinusoidal and non-sinusoidal signals.

PEP-to-average SSB is 2.5 to 1,

PEP-to-average depends on speech characteristics.

Very high speed allows digitizing high frequencies.

Analog-to-digital converter with 8 bit resolution can encode 2^8 = 256 levels.

Low-pass filter removes spurious sampling artifacts.

SIGNALS AND EMISSIONS

Analog-to-digital conversion quality measured by total harmonic distortion.

GOURP 8B – MODULATION AND DEMODULATION

Modulation index is ratio of frequency deviation to modulating signal frequency.

Modulation index of phase-modulated emission does not depend on frequency.

Modulation index with max deviation of 3000 Hz and modulating frequency of 1000 Hz is 3.

With deviation of 6 kHz and 2 kHz modulating frequency, index is 3.

Deviation ratio is ratio of maximum carrier frequency deviation to highest audio modulating frequency.

Maximum frequency swing of 5 kHz and maximum modulating frequency of 3 kHz is 1.67 deviation ratio,

Maximum frequency swing of 7.5 kHz and maximum modulating frequency of 3.5 kHz is 2.14 deviation ratio.

Orthogonal frequency division uses subcarriers at frequencies chosen to avoid interference.

Frequency division multiplexing is dividing the signal in separate frequency bands.

Digital time division multiplexing uses discrete time slots.

GROUP 8C – DIGITAL SIGNALS

Quadrature Amplitude Modulation uses two signals 90 degrees out of phase.

Symbol rate is rate waveform changes.

PSK phase-shifting should be done at zero crossings to minimize bandwidth.

Minimize bandwidth of PSK using sinusoidal data pulses.

Bandwidth of 13-WPM Morse is 52 Hz.

Bandwidth of FT8 is 50 Hz.

Bandwidth of ASCII FM is 15.36 kHz.

ARQ error correction requests retransmission .

Gray code.

Increase data rate without increasing bandwidth by using a more efficient mode.

Symbol rate and baud rate are the same.

CW bandwidth set by keying speed and shape factor.

Constellation diagram is phase and amplitude states for each symbol.

Mesh network nodes have IP addresses.

Mesh networks form using discovery and link establishment protocols.

GROUP 8D – KEYING DEFECTS

Signals not following a spread spectrum algorithm are suppressed.

Spread spectrum uses direct sequence.

Spread spectrum uses a pseudorandom sequence.

Short rise and fall times generate key clicks.

Reduce key clicks by increasing rise and fall times.

Use a parity bit in an ACII stream to detect errors.

Excessive transmit audio levels can cause overmodulation.

IMD is measure of distortion caused by excessive input audio.

PSK IMD -30 dB.

Baudot uses 5 data bits per character.

ASCII   transmits both upper and lower case.

E9 – ANTENNAS AND TRANSMISSION LINES

E9A Basic antenna parameters

An isotropic radiator is a hypothetical, lossless antenna having *equal radiation intensity in all directions* used as a reference for antenna gain. *Hint: The word "isotropic" means "equal way." An isotropic antenna radiates equally in all directions. That is all you need to recognize.*

The term that describes total radiated power, taking into account all gains and losses is *effective radiated power.* You determine the effective radiated power of a system by adding and subtracting the dB loss and gain of each component.

The effective radiated power (ERP) of a repeater station with 150 watts transmitter power output, 2 dB feed line loss, 2.2 dB duplexer loss, and 7 dB antenna gain is 286 watts. *Solve: 7 -2 -2.2 = 2.8 dB gain. 3 dB gain is double, so the answer is the one a little less than double. The other answers are way off.*

The effective radiated power relative to a dipole of a station with 200 watts transmitter power, 4 dB feed line loss, 3.2dB duplexer loss and 10 dBd of antenna gain is *317 watts.* *Solve: First the dBs: 10 dB gain -4 -3.8 = 2.2 dB. 3 dB of gain is double the power or 400 watts. The answer is the one that is a little less than 400 = 317 watts. Don't memorize the number. It is the only one a little less than double.*

The effective radiated power of a repeater station with 200 watts transmitter power, 2 dB

feed line loss, 2.8 dB duplexer loss, 1.2 dB circulator loss and a 7 dBi antenna gain is 252 watts. *Solve: The total gain is 7 – 2 – 2.8 – 1.2 = 1 dB. The answer is the one just over 200 watts which is also the only answer above 200.*

The factor that may affect the feed point impedance of an antenna is the *antenna height*. Height influences impedance because the antenna capacitively couples to the ground. The higher the antenna, the less the effect.

Ground gain is *increase in signal strength from ground reflections*.

The band with the smallest first Fresnel zone is *5.8 GHz*. The Fresnel zone describes a ring around the transmitted signal. The higher the frequency, the shorter the wavelength and the smaller the ring.

Antenna efficiency is *radiation resistance divided by total resistance*. *Hint: The radiation resistance is what is radiating. The total resistance is everything, including the loss resistance. The more radiating, the better.*

The factor that determines ground losses for a ground-mounted vertical antenna operating on HF is *soil conductivity*. *Hint: Soil is not very conductive and therefore increases the losses.*

A way to improve the efficiency of a ground-mounted quarter-wave vertical antenna is to *install a ground radial system*. *Hint: The radials provide the other half of the dipole and reduce ground losses.*

A half-wavelength dipole has about 2.15 dB gain over an isotropic antenna. The following question asks you

to compare the two. Subtract 2.15 from the gain over the isotropic antenna to get the dipole gain.
If an antenna has 6 dB gain over an isotropic radiator, it is only *3.85 dB* better than a half-wavelength dipole *Solve: 6 – 2.15 = 3.85.*

E9B Antenna Patterns

The beam-width for a directional antenna is the two points where the signal strength of the antenna is 3 dB less than maximum (half).

Figure E9-1, below, is an antenna azimuth chart showing the pattern's strength in various directions.
The beam-width in the antenna radiation pattern shown in Figure E9-1 is *50 degrees*. *Solve: The pattern hits the -3 dB ring at a little over 30 degrees and less than -30 degrees, so the width is a little less than the total of 60 degrees. 50 degrees is the closest answer.*

Figure E9-1

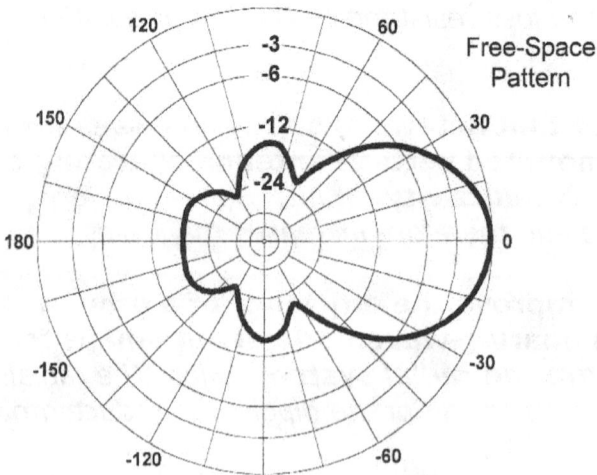

The front-to-back ratio is *18 dB*. *Solve: The front is at 0 dB, the back is at -18 dB (not well marked but it is halfway between the -12 and -24 circles)*

The front-to-side ratio is *-14 dB*. *Solve: The front is 0 dB, and the side is -14 dB (again, not well marked but if you look carefully, you see it is closer to the -12 circle).*

The next questions refer to Figure 9-2.

Figure E9-2

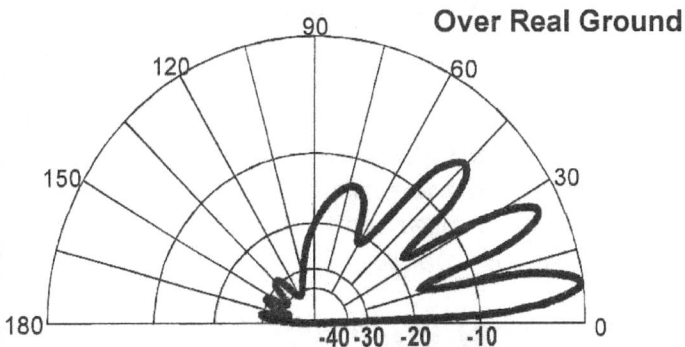

Over Real Ground

In Figure E9-2 the antenna pattern is an *elevation pattern*. It shows the takeoff angles.

The front-to-back ratio shown in Figure E9-2 is *28 dB*. It isn't well marked but you can see the back is just a little over the -30 dB arc.

The elevation angle of the peak response is *7.5 degrees*. The strongest lobe is the bottom at 7.5 degrees.

The difference in radiated power between a lossless antenna with gain and an isotropic radiator driven by the same power *is the same*. *Hint: The total amount of radiation is the same. The directional antenna concentrates it.*

ANTENNAS AND TRANSMISSION LINES

The far field of an antenna is the region where the shape of the antenna pattern *no longer varies with distance.* *Hint: Far enough away that ground or objects no longer affect the pattern.*

Antenna modeling programs use a computer program technique called *Method of Moments.*

The principle of a Method of Moments analysis is a wire is modeled as a series of segments, each having a *uniform value of current.* *Hint: Current causes the antenna to radiate. Uniform current segments are easier to calculate.*

The disadvantage of decreasing the number of wire segments in an antenna model below 10 segments per half-wavelength is *the computed feed point impedance may be incorrect.* *Hint: Fewer data points mean less precision. You don't need to know how many.*

E9C Practical Wire Antennas

The radiation pattern of two 1/4-wavelength vertical antennas spaced 1/2-wavelength apart and fed 180 degrees out of phase is a *figure-8 oriented along the axis of the array.*

The radiation pattern of two 1/4-wavelength vertical antennas spaced 1/4- wavelength apart and fed 90 degrees out of phase is *cardioid*.

The radiation pattern of two 1/4-wavelength vertical antennas spaced 1/2-wavelength apart and fed in phase is a *figure-8 broadside* to the axis of the array.
Memory point:
180 out of phase, oriented along.
90 out of phase, cardioid.
In phase, broadside.

As the wire length is increased for an unterminated long wire antenna, the radiation pattern changes as *more lobes form increasingly aligned with the axis of the antenna*. *Hint: Longer wire is more lobes.*

An off-center-fed-dipole (OCFD) antenna is fed between the center and one end to *create a similar feed point impedance on multiple bands*. *Hint: OCFD is a multi-band antenna. That is all you need to recognize to answer the question.*

A rhombic antenna is two long wires pointed in the same direction in the shape of a rhombus. **The effect of a terminating resistor on a rhombic antenna or long wire is to *change the radiation pattern from bidirectional to unidirectional*.**

A folded dipole antenna is a half-wave dipole with an *additional parallel wire connecting its two ends*. *Hint: It folds back on itself.*

A two-wire folded dipole antenna has a feed point impedance at the center of *300 ohms*.

A G5RV is a wire antenna center-fed through a specific length of *open-wire line* connected to a balun and coaxial feed line. *Hint: the G5RV uses open-wire line. That is all you need to know.*

A Zepp antenna is an *end-fed* half-wavelength dipole. *Hint: Zepp antennas were an end-fed wire trailing behind a Zeppelin.*

The far-field elevation pattern of a vertically polarized antenna mounted over seawater versus soil is *radiation at low angles increases*. *Hint: Far-off DX signals arrive at a low angle. Verticals over saltwater are great low-angle DX antennas.*

ANTENNAS AND TRANSMISSION LINES

An extended double Zepp antenna is a *center fed 1.25 wavelength* antenna. *Hint: It is extended, so it is extra long. This Zepp is not end fed.*

The radiation pattern of a horizontally polarized antenna varies with increasing height in that the *takeoff angle decreases*. *Hint: The signal starts higher and therefore travels further before it hits the ground. It bounces off the ground at a lower angle.*

A horizontally polarized antenna mounted on a long slope compared to one mounted on flat ground will differ in that the *takeoff angle decreases in the downhill direction*. *Hint: The signal bounces off the hill further away and at a lower elevation.*

E9D *Directional and Short Antennas*

When the operating frequency is doubled, the gain of an ideal parabolic dish is increased by *6 dB*. *Solve: Doubling the frequency increases the gain by $2^2 = 4$ times or 6 dB. Careful! The answer is not 4 dB.*

Two linearly polarized Yagi antennas can be used to produce circular polarization by arranging the 2 Yagis on the same axis *perpendicular to each other* with the driven elements at the same point on the boom fed 90 degrees out of phase. *Hint: Arrange the two yagis perpendicular to each other. Forget the rest.*

Coils are often used to lengthen an electrically short antenna. **The most efficient location for a loading coil on an electrically short whip is *near the center of the vertical radiator*.**

Antenna loading coils should have a high ratio of reactance to resistance *to maximize efficiency*.

Hint: Higher resistance would mean higher losses and lower efficiency.

A Yagi's driven element is *1/2 wavelength long.*
Hint: A Yagi starts as a dipole.

When one or more loading coils are used to resonate an electrically short antenna, the SWR *bandwidth is decreased*. Shortened antennas can have very narrow bandwidth.

An advantage of top loading an electrically short HF antenna is *improved radiation efficiency*
Hint: The maximum current is at the feed point (base). Let the antenna radiate before it suffers losses in a coil or loading.

As the Q of an antenna increases, the SWR *bandwidth decreases*. *Hint: Higher Q means sharper tuning and bandwidth decreases.*

The function of a loading coil used as part of an HF mobile antenna is to *resonate the antenna* by canceling the capacitive reactance. *Hint: Resonate the antenna.*

The radiation resistance of a base-fed whip antenna when operated below its resonant frequency *decreases*. *Hint: The antenna becomes less efficient – the radiation resistance decreases.*

Most two-element Yagis with normal spacing have a reflector instead of a director for *higher gain*. Of the two, the reflector contributes more gain.

The purpose of making a Yagi's parasitic elements either longer or shorter than resonance is *control of phase shift*.

E9E Impedance Matching

The matching system for Yagi antennas that requires the driven element be insulated from the boom is *beta or hairpin.*

The antenna matching system that matches an coaxial cable to an antenna by connecting the shield to the center of the antenna and the conductor a fraction of a wavelength to one side is called a *gamma match.* *Hint: Fed in the center and another place is gamma.*

An effective method of shunt-feeding a grounded tower at its base is a *gamma match.* *Hint: Another use for the gamma match.*

~~**The matching system that places an inductance across the feed point of a vertical monopole antenna is *beta or hairpin.***~~ —Deleted

The matching system that uses a section of transmission line connected in parallel with the feed line at or near the feed point is called *stub match*. *Hint: The parallel feed line is a stub.*

Memory points:
Insulated from the boom: Beta or hairpin
Fraction of a wavelength to the side: Gamma match
Shunt feed a tower: Gamma match
Parallel with the feedline: Stub match

The purpose of a series capacitor in a gamma-match is to *cancel unwanted inductive reactance.* *Hint: Capacitors cancel inductance.*

The Yagi driven element tuned to use a hairpin matching system, should be *capacitive* (shorter than 1/2 wavelength. *Hint: Hairpins are short.*

A transmission line suitable for constructing a quarter-wave Q-section for matching a 100-ohm feed point impedance to a 50-ohm coaxial cable feed would be **75 ohms**.   A quarter-wave of 75 ohm cable transforms 100 ohms to 50 ohms.

The term used to describe the interaction at the load and transmission line is **reflection coefficient.** *Hint: A line will reflect power back.*

A use for a Wilkinson divider is to **divide power** equally between two 50-ohm loads while maintaining 50-ohm input impedance. *Hint: It is a divider, so it divides power. The other answer about dividing frequency doesn't make any sense.*

The purpose of using multiple driven elements connected through phasing lines is **to control the antenna's radiation pattern.** *Hint: Phasing lines control the pattern.*

E9F Transmission Lines

Wave velocity is less in a transmission line than in free space. **The velocity factor of a transmission line is the velocity of the wave in the transmission line *divided by the velocity of light in a vacuum.*** *Hint: It is the ratio of speed in the line compared to the speed of light in a vacuum.*

The velocity factor of a transmission line is determined by the **insulating dielectric material used in the line.**   The type of insulation.

The electrical length of a coaxial cable is longer than its physical length because **electromagnetic waves move more slowly** in a coaxial cable than in air.   *Hint: It takes the wave longer to travel so the electrical length is longer.*

ANTENNAS AND TRANSMISSION LINES

Microstrip is *precision printed circuit conductors* above a ground plane that provide constant impedance interconnects at microwave frequencies. *Cheat: Pick one of those attributes to recognize and ditch the over-explanation.*

When cutting line for a certain frequency, take the velocity factor into account. The velocity factor for polyethylene dielectric coax is about .67. So cut it to 67% of the calculated length, and the wave will arrive at the right time.

The approximate physical length of an air-insulated parallel transmission line that is electrically one-half wavelength long at 14.1 MHz would be *10.6 meters*.
Hint: Air-insulated line has no dielectric, so the velocity factor is very close to 1.
Solve: The frequency of 14.1 MHz would have a wavelength of 300/14.1 = 21.28 meters, and one-half wavelength would be 10.6 meters.

Parallel conductor transmission line *has lower loss* than coaxial cable with a plastic dielectric.
Hint: Ladder line has much lower losses than coax at all frequencies.

The significant difference between foam dielectric coaxial cable and solid dielectric cable is:
- **Foam has lower safe operating voltage limits.**
- **Foam has lower loss.**
- **Foam has higher velocity factor.**
- **All these choices are correct.**

Transmission line can act as a transformer. A half-wave line mirrors the end impedance. **The impedance of a 1/2 wavelength transmission** line when the line is shorted at the far end is a **very low impedance.** *Hint: Same as the end.*

A quarter-wave transmission line inverts the impedance. **The impedance of a 1/4-wavelength line when the line is open at the far end is very low impedance.** *Hint: Opposite.*

The impedance of a 1/4- wavelength transmission line when the line is shorted at the far end is very high impedance. *Hint: Opposite.*

An open or shorted 1/8-wavelength transmission line results in a capacitive or inductive reactance. **The impedance of a 1/8-wavelength transmission line when the line is open at the far end is capacitive reactance.** *Cheat: Think of the center and shield as parallel plates of a capacitor.*

The impedance of a 1/8 wave transmission line when the line is shorted at the far end is an inductive reactance. *Cheat: The opposite of above.*

Memory Points:
1/2 wave mirrors
1/4 wave inverts
1/8 wave open capacitor
1/8 wave shorted inductor

E9G Smith Chart

Using a Smith Chart, you can calculate impedance along transmission lines. *Hint: The impedance seen at the feed point changes along the length of a transmission line. The Smith Chart is a way of computing without using incredibly complex math.*

The coordinate system on a Smith Chart is resistance circles and reactance arcs. *Hint:*

ANTENNAS AND TRANSMISSION LINES

Impedance is resistance and reactance. The chart is a circle. Resistance circles.

A Smith Chart is used to determine *impedance and SWR values in transmission lines.*

The two families of circles that make up a Smith chart are *resistance and reactance.*

A common use for the Smith chart is to determine the *length and position of an impedance matching stub*.

The name of the large outer circle on which the reactance arcs terminate is the *reactance axis*. *Hint: It is showing the value of the reactance on that arc. Reactance arcs relate to the reactance axis.*

Figure E9-3

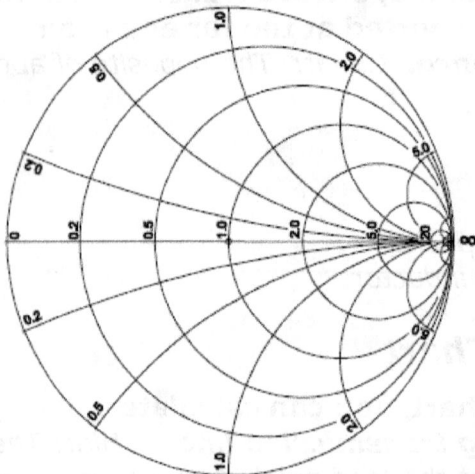

The chart shown in E9-3 is a Smith Chart.

The only straight line shown is called the *resistance axis*. *Hint: Resistance doesn't change with frequency.*

The arcs on a Smith Chart represent points with a **constant reactance**. *Hint: Arcs are reactance.*

Normalize a Smith Chart by **reassigning the prime center's impedance values.** *Hint: You make it normal by reassigning to the prime center.*

The third family of circles often added to a Smith Chart during the process of designing impedance matching networks are **constant-SWR circles.**

The wavelength scales on a Smith Chart are calibrated in **fractions of transmission line electrical wavelength.** *Hint: You are calculating a transmission line depending on its length.*

E9H Receiving Antennas

When constructing a Beverage antenna, it should be **at least one wavelength long**. *Hint: Beverages are long and low. They minimize noise on receive.*

For a 160-meters and 80-meter receiving antenna, the atmospheric noise is so high that **directivity is much more important than losses.** *Hint: In fact, gain might make the noise worse.*

Receiving directivity factor (RDF) is **peak antenna gain compared to average gain** over the hemisphere around and above the antenna *Hint: Peak to average.*

The purpose of placing an electrostatic shield around a small wire-loop antenna for direction finding is it eliminates unbalanced capacitive coupling to the surroundings, **improving the depth of its nulls.** *Hint: When using a direction-finding antenna, turn it to find the null in the signal because a null is easier to discern than a peak.*

ANTENNAS AND TRANSMISSION LINES

"Improving the nulls" is an advantage and all you need recognize to answer.

A small loop antenna will have a sharp null. **The challenge of a small wire-loop antenna for direction finding is that it is *bidirectional*.** *Hint: It is a drawback to hear nulls in both front and back.*

The correct value for a Beverage antenna's terminating resistor is one that has *minimum variation in SWR over the desired frequency range*.

The function of a Beverage antenna's termination resistor is to *absorb signals from the reverse direction*. Making the antenna one directional.

The function of a sense antenna is it modifies the pattern of DF antenna array to *provide a null in one direction*. *Hint: A null in one direction is preferable to two.*

The radiation pattern created by a single-turn, terminated loop such as a pennant antenna is *cardioid*.

The output voltage of a multiple-turn receiving loop antenna can be increased by *increasing the number of turns and/or the area enclosed by the loop*. *Hint: Bigger is better.*

The feature of a cardioid pattern antenna that makes it useful for direction finding is a *single null*. *Hint: A single null gives the best directivity.*

SUMMARY: ANTENNAS AND TRANSMISSION LINES

GROUP 9A – BASIC ANTENNA PARAMETERS

Isotropic is equal radiation in all directions.

Total radiated power is effective radiated power.

Calculate effective radiated power by adding gains and losses.

Feedpoint impedance affected by antenna height.

Ground gain is signal strength increase from ground reflections.

Smallest Fresnel zone on 5.8 GHz.

Antenna efficiency is radiation resistance divided by total resistance.

Ground losses determined by soil conductivity.

To improve efficiency of ground mounted vertical, install radials.

Gain of dipole vs isotropic, subtract 2.15 dB .

GROUP 9B – ANTENNA PATTERNS

Beam width is between two 3 dB points on graph.

Figure is an elevation pattern.

Radiated power in a lossless antenna is the same as an isotropic.

Far field is where pattern no longer varies with distance.

Antenna modeling uses Method of Moments.

With Method of Moments, each segment has uniform value of current

The computed impedance may be incorrect if you decrease the number of wire segments.

GROUP E9C – WIRE ANTENNAS

If two verticals are fed:

180 degrees out of phase – oriented along axis

90 degrees out of phase - cardioid

ANTENNAS AND TRANSMISSION LINES

Increasing antenna length forms additional lobes.

Off-center fed dipole creates same impedance on multiple bands.

On rhombic, a terminating resistor changes antenna from bi-directional to uni-directional.

Folded dipole has additional parallel wire connecting two ends.

Folded dipole has 300-ohm feedpoint impedance.

G5RV is fed with specific length of open wire line.

Zepp is end-fed half-wavelength dipole.

Vertical over seawater increases low-angle radiation.

Extended double Zepp is 1.25 wavelengths long.

Horizontal takeoff angle decreases as antenna raised.

On a low slope, takeoff angle decreases in downhill direction.

GROUP E9D – DIRECTIONAL AND SHORT

Doubling operating frequency increases parabolic dish's gain by 6 dB.

Produce circular polarization by mounting yagis perpendicular on same axis.

Most efficient location for coil is center of vertical radiator.

Loading coil should have high ratio of reactance to resistance.

Yagi's driven element is ½ wavelength long.

More loading coils reduce SWR bandwidth.

Top-loading improves radiation efficiency.

A Q increases, SWR bandwidth decreases.

Loading coil resonated antenna by canceling capacitive reactance.

Radiation resistance decreases when an antenna operated below its resonant frequency.

2-element Yagi's use reflector not director for higher gain.

Yagi's elements are shorter or longer to control phase shift.

GROUP 9E – IMPEDANCE MATCHING

Yagi insulated boom is beta or harpin matching .

Feed to the side is gamma match.

Shunt feed a tower with gamma match.

Inductance across feed point is beta or hairpin.

Parallel section of transmission line is stub match.

Series capacitor is to cancel inductive reactance.

Yagi driven element should be capacitive to use hairpin (shorter than 1/2 wavelength).

Quarter-wave section to match 100 ohm and 50 ohm should be 75 ohms.

Reflection coefficient is interaction of load and line.

Wilkinson divider divides power equally.

Use phasing lines to control the radiation pattern.

GROUP E9F – TRANSMISSION LINES

Velocity factor is speed in line divided by speed of light.

Velocity factor determined by insulating dielectric material. In the line.

Electrical length of coax is longer because waves move slower in the cable than in air.

Microstrip is precision printed conductors.

Air-insulated transmission line has velocity factor close to 1. 1/4 wave would be 10.6 meters.

Parallel conductor transmission line has lower loss than coax.

Foam dielectric has lower safe operating voltages, lower loss, and higher velocity factor.

As a transmission line transformer:
1/2 wave mirrors the end
1/4 wave inverts the end
1/8 wave open capacitor
1/8 wave shorted inductor.

ANTENNAS AND TRANSMISSION LINES

GROUP E9G – SMITH CHART

Calculates impedance along transmission lines.

Coordinate system is resistance circles and reactance arcs.

Used to determine impedance and SWR in transmission lines.

Determine length and position of impedance matching stub.

Name of large outer circle is the reactance axis.

Only straight line is the resistance axis.

Arcs are points of constant reactance.

Normalize by reassigning the prime center.

Third family of circles is constant-SWR circles.

Wavelength scales calibrated in fractions of wavelength.

GROUP E9H – RECEIVING ANTENNAS

Beverage should be at least one wavelength long.

Atmospheric noise is so high directivity is more important than gain.

Receiving diversity factor is peak gain compared to average gain.

Electrostatic shield on loop antenna improves depths of nulls.

Small loop antenna is bidirectional.

Beverage terminating resistor should have minimum variation on SWR.

Terminating resistor absorbs signals from the reverse.

Sense antenna provides null in one direction.

Pennant antenna has cardioid pattern.

Increase output by increasing number of turns.

Cardioid pattern is good for direction finding because it has single null.

E0 – SAFETY

E0A Safety

The primary function of an external earth connection or ground rod is *lightning charge dissipation*. *Hint: You want to dissipate the lightning force outside before it gets in the house.*

MPE limits are the Maximum Permissible Exposure to electromagnetic radiation. **When evaluating RF exposure levels from your station to your neighbor's house, you must make sure the signals from your station are *less than the uncontrolled Maximum Permitted Exposure limits.*** *Hint: Your neighbor can't control what you are doing, so he is entitled to the protection of the lower, uncontrolled exposure limits. You are evaluating exposure levels, not emission levels.*

The range of frequencies over which the FCC human body exposure limits are the most restrictive are *30 – 300 MHz*. *Hint: VHF, not UHF, is the most dangerous. Very High Fry-ability.*

When evaluating a site with multiple transmitters operating at the same time, the operators and licensees of transmitters responsible for mitigating over-exposure are each transmitter that produces *5% or more* of its MPE limit in areas where the total MPE is exceeded. *Hint: Very confusing answer. Look for the answer with 5%. Practically everyone is responsible for mitigating an unsafe situation.*

A potential hazard of using microwaves in the amateur radio bands is the *high gain antennas commonly used can result in high exposure levels*. *Hint: Microwaves heat.*

SUMMARY

There are separate electric (E) and magnetic (H) field MPE limits because:
- **The body reacts to electromagnetic radiation from both fields.**
- **Ground reflections and scattering make the field strength vary with location.**
- **E field and H field radiation intensity peaks can occur at different locations.**
- **All these choices are correct.**

SAR measures the rate at which RF energy is **absorbed by the body**. *Hint: SAR means Specific Absorption Rate or "Soon about to roast."*

Equipment exempt from RF exposure evaluations are **hand-held transceivers sold before May 3, 2021.** *Hint: They didn't make the requirements retroactive.*

RF exposure evaluation must be performed on 80 meters **always.** Other than the question above, there are no other exemptions.

Regarding tower safety, "100% tie-off " means **at least one lanyard attached to the tower at all times.**

When climbing, lanyards should be attached **to tower legs.** *Hint: They are the strongest part of the tower.*

A shock-absorbing lanyard should be attached to a tower **above the climber's head level.**

SUMMARY: SAFETY

GROUP 0A – SAFETY

Purpose of external ground rod is lightning protection.

Make sure neighbor is exposed to less than uncontrolled Maximum Permitted Exposure.

Exposure limits most restrictive on 3-300 MHz.

In multi-transmitter, everyone over 5% must mitigate.

Hazard of microwaves is high gain antennas.

Separate electric and magnetic MPE limits because body reacts to both, ground reflection varies, peaks occur at different locations.

SAR measures the rate RF energy is absorbed by the body. (Soon About to Roast).

Equipment exempt are hand-helds sold before May 3, 2021.

Evaluation must be performed on 80 meters always.

100% tie-off is at least one lanyard attached at all times.

Attach lanyard to tower legs.

Shock absorbing lanyard above the climber's head level.

AMATEUR RADIO EXTRA CLASS QUICK REVIEW

NOTE: The question pool is divided into Groups within Subelements. There is one question from each of the 50 Groups. If you have trouble with a Group, don't worry. You can only get one question per Group, and you may know the answer to the one.
The Figures (diagrams) are at the back of the book.

E1 — COMMISSION RULES

[6 Exam Questions — 6 Groups]

E1A   Frequency Priviledges

E1A01 Why is it not legal to transmit a 3 kHz bandwidth USB signal with a carrier frequency of 14.348 MHz?
The upper 1 kHz of the signal is outside the 20-meter band

E1A02 When using a transceiver that displays the carrier frequency of phone signals, What displayed frequencies represents the lowest frequency at which a properly adjusted LSB emission will be totally within the band?
3 kHz above the lower band edge

E1A03 What is the highest legal carrier frequency on the 20-meter band for transmitting a 2.8 kHz wide USB data signal?
14.1472 MHz

E1A04 May an Extra class operator answer the CQ of a station on 3.601 MHz LSB phone?
No, the sideband components will extend beyond the edge of the phone band segment

E1A05 Who must be in physical control of the station apparatus of an amateur station aboard any vessel or craft that is documented or registered in the United States?

Any person holding an FCC issued amateur license or who is authorized for alien reciprocal operation

E1A06 What is the required transmit frequency of a CW signal for channelized 60 meter operation?
At the center frequency of the channel

E1A07 What is the maximum power permitted on the 2200 meter band?
1 watt EIRP (Equivalent isotropic radiated power)

E1A08 If a station in a message forwarding system inadvertently forwards a message that is in violation of FCC rules, who is primarily accountable for the rules violation?
The control operator of the originating station

E1A09 Except in some parts of Alaska, what is the maximum power permitted on the 630-meter band?
5 watts EIRP (equivalent isotropic radiated power)

E1A10 If an amateur station is installed aboard a ship or aircraft, before the station is operated?
Its operation must be approved by the master of the ship or the pilot in command of the aircraft

E1A11 What licensing is required when operating an amateur station aboard a US-registered vessel in international waters?
Any FCC-issued amateur license

E1B Station Restrictions

E1B01 What constitutes a spurious emission?
An emission outside the signal's necessary bandwidth that can be reduced or eliminated without affecting the information transmitted

E1B02 An acceptable bandwidth for digital voice or slow-scan TV transmissions made on the HF amateur bands?
3 kHz

E1B03 Within what distance must an amateur station protect an FCC monitoring facility from harmful interference?
1 mile

E1B04 What must the control operator of a repeater operating in the 70-centimeter band do if a radiolocation system experiences interference from that repeater?
Cease operation or make changes to the repeater that mitigate the interference

E1B05 What is the National Radio Quiet Zone?
An area surrounding the National Radio Astronomy Observatory

E1B06 What additional rules apply if you are installing an amateur station antenna at a site at or near a public use airport?
You may have to notify the Federal Aviation Administration and register it with the FCC as required by Part 17 of the FCC rules

E1B07 To what type of regulations does PRB-1 apply?
State and local zoning

E1B08 What limitations may the FCC place on an amateur station if its signal causes interference to domestic broadcast reception, assuming that the receivers involved are of good engineering design?
The amateur station must avoid transmitting during certain hours on frequencies that cause the interference

E1B09 Which amateur stations may be operated under RACES rules?
Any FCC-licensed amateur station certified by the responsible civil defense organization for the area served

E1B10 What frequencies are authorized to an amateur station operating under RACES rules?
All amateur service frequencies authorized to the control operator

REVIEW – COMMISSION RULES

E1B11 What does PRB-1 require of regulations affecting amateur radio?
Reasonable accommodations of amateur radio must be made

E1C Control

E1C01 What is the maximum bandwidth for a data emission on 60 meters?
2.8 kHz

E1C02 What types of communications may be transmitted to amateur stations in foreign countries?
Communications incidental to the purpose of the amateur service and remarks of a personal nature

E1C03 How long must an operator wait after filing a notification with the Utilities Technology Council (UTC) before operating on the 2200-meter or 630-meter band?
Operators may operate after 30 days, providing they have not been told that their station is within 1 kilometer of PLC systems using those frequencies

E1C04 What is meant by IARP?
An international amateur radio permit that allows US amateurs to operate in certain countries of the Americas

E1C05 Under what situation may a station transmit third party communications while being automatically controlled?
Only when transmitting RTTY or data emissions

E1C06 What is required in order to operate in accordance with CEPT rules in foreign countries where permitted?
You must have a copy of FCC Public Notice DA 16-1048

E1C07 What notifications must be given before transmitting on the 630- or 2200-meter bands?
Operators must inform the Utilities Technology Council (UTC) of their call sign and coordinates of the station

E1C08 What is the maximum permissible duration of a remotely controlled station's transmissions if its control link malfunctions?
3 minutes

E1C09 What is the highest modulation index permitted at the highest modulation frequency for angle modulation below 29.0 MHz?
1.0

E1C10 What is the permitted mean power of any spurious emission below 30 MHz with respect to the mean power of the fundamental emission?
- 43 dB below

E1C11 What operating arrangements allows an FCC-licensed US citizen to operate in many European countries, and alien amateurs from many European countries to operate in the US?
CEPT agreement

E1C12 On what portion of the 630 meter band are phone emissions permitted?
The entire band

E1D Amateur Space and Earth Stations

E1D01 What is the definition of telemetry?
One-way transmission of measurements at a distance from the measuring instrument

E1D02 What may transmit encrypted messages?
Telecommand signals from a space telecommand station

E1D03 What is a space telecommand station?
An amateur station that transmits communications to initiate, modify, or terminate functions of a space station

E1D04 What is required in the identification transmissions from a balloon-borne telemetry station?
Call sign

E1D05 What must be posted at the station location of a station being operated by telecommand on or within 50 km of the earth's surface?
A photocopy of the station license
A label with the name, address, and telephone number of the station licensee
A label with the name, address, and telephone number of the control operator
All these choices are correct

E1D06 What is the maximum permitted transmitter output power when operating a model craft by telecommand?
1 watt

E1D07 Which HF amateur bands have frequencies authorized for space stations?
Only the 40, 20, 15, and 10 meters

E1D08 Which VHF amateur bands have frequencies authorized for space stations?
2 meters

E1D09 Which UHF amateur bands have frequencies authorized for space stations?
70 cm and 13 cm

E1D10 Which amateur stations are eligible to be telecommand stations of space stations, subject to the privileges of the class of operator license held by the control operator of the station?
Any amateur station so designated by the space station licensee
E1D11 Which amateur stations are eligible to operate as Earth stations?

Any amateur station, subject to the privileges of the class of operator license held by the control operator.

E1D12 What amateur stations may transmit one-way communications?
A space station, beacon station, or telecommand station

E1E Examiners

E1E01 For which types of out-of-pocket expenses do the Part 97 rules state that VEs and VECs may be reimbursed?
Preparing, processing, administering, and coordinating an examination for an amateur radio operator license

E1E02 Who does Part 97 task with maintaining the pools of questions for all US amateur license examinations?
The VECs

E1E03 What is a Volunteer Examiner Coordinator?
An organization that has entered into an agreement with the FCC to coordinate, prepare, and administer amateur operator license examinations

E1E04 What is required to be accredited as a Volunteer Examiner?
The VEC must confirm that the VE applicant meets FCC requirements to serve as an examiner

E1E05 What must the VE team do with the application form if the examinee does not pass the exam?
Return the application document to the examinee

E1E06 Who is responsible for the proper conduct and necessary supervision during an amateur operator license examination session?
Each administering VE

E1E07 What should a VE do if a candidate fails to comply with the examiner's instructions during an amateur operator license examination?
Immediately terminate the candidate's examination

E1E08 To What examinees may a VE not administer an examination?
Relatives of the VE as listed in the FCC rules

E1E09 What may be the penalty for a VE who fraudulently administers or certifies an examination?
Revocation of the VE's amateur station license grant and the suspension of the VE's amateur operator license grant

E1E10 What must the administering VEs do after the administration of a successful examination for an amateur operator license?
They must submit the application document to the coordinating VEC according to the coordinating VEC instructions

E1E11 What must the VE team do if an examinee scores a passing grade on all examination elements needed for an upgrade or new license?
Three VEs must certify that the examinee is qualified for the license grant and that they have complied with the administering VE requirements

E1F Miscellaneous Rules

E1F01 On what frequencies are spread spectrum transmissions permitted?
Only on amateur frequencies above 222 MHz

E1F02 What privileges are authorized in the US to persons holding an amateur service license granted by the government of Canada?

REVIEW – COMMISSION RULES

The operating terms and conditions of the Canadian amateur service license, not to exceed US Amateur Extra Class license privileges

E1F03 Under what circumstances may a dealer sell an external RF power amplifier capable of operation below 144 MHz if it has not been granted FCC certification?
It was constructed or modified by an amateur operator for use at an amateur station

E1F04 What geographic descriptions approximately describes "Line A"?
A line roughly parallel to and south of the border between the US and Canada

E1F05 Amateur stations may not transmit in What frequency segments if they are located in the contiguous 48 states and north of Line A?
420 MHz - 430 MHz

E1F06 Under what circumstances might the FCC issue a Special Temporary Authority (STA) to an amateur station?
To provide for experimental amateur communications

E1F07 When may an amateur station send a message to a business?
When neither the amateur nor his or her employer has a pecuniary interest in the communications

E1F08 What types of amateur station communications are prohibited?
Communications transmitted for hire or material compensation, except as otherwise provided in the rules

E1F09 What cannot be transmitted over an amateur radio mesh network?
Messages encoded to obscure their meaning

E1F10 Who may be the control operator of an auxiliary station?
Only Technician, General, Advanced or Amateur Extra Class operators

E1F11 What best describes one of the standards that must be met by an external RF power amplifier if it is to qualify for a grant of FCC certification?
It must satisfy the FCC's spurious emission standards when operated at the lesser of 1500 watts or its full output power

E2 - OPERATING PROCEDURES

[5 Exam Questions - 5 Groups]

E2A Amateur Radio in Space

E2A01 What is the direction of an ascending pass for an amateur satellite?
From south to north

E2A02 What is characteristic of an inverting linear transponder?
Doppler shift is reduced because the uplink and downlink shifts are in opposite directions
Signal position in the band is reversed
Upper sideband on the uplink becomes lower sideband on the downlink, and vice versa
All these choices are correct

E2A03 How is the signal inverted by an inverting linear transponder?
The signal is passed through a mixer and the difference product is transmitted

E2A04 What is meant by the term "mode" as applied to an amateur radio satellite?
The satellite's uplink and downlink frequency bands

E2A05 What do the letters in a satellite's mode designator specify?
The uplink and downlink frequency ranges

E2A06 What are Keplerian elements?
Parameters that define the orbit of a satellite

E2A07 What types of signals can be relayed through a linear transponder?
FM and CW
SSB and SSTV

PSK and packet
All these choices are correct

E2A08 Why should effective radiated power to a satellite that uses a linear transponder be limited?
To avoid reducing the downlink power to all other users

E2A09 What do the terms "L band" and "S band" specify regarding satellite communications?
The 23-centimeter and 13-centimeter bands

E2A10 What type of satellite appears to stay in one position in the sky?
Geostationary

E2A11 What type of antenna can be used to minimize the effects of spin modulation and Faraday rotation?
A circularly polarized antenna

E2A12 What is the purpose of digital store-and-forward functions on an amateur radio satellite?
To store digital messages in the satellite for later download

E2A13 What techniques is normally used by low Earth orbiting digital satellites to relay messages?
Store-and-forward

E2B Television Practices

E2B01 In digital television, a coding rate of 3/4 means
25% of the data sent is forward error correction data

E2B02 How many horizontal lines make up a fast-scan (NTSC) television frame?
525

E2B03 How is an interlaced scanning pattern generated in a fast-scan (NTSC) television system?

REVIEW – OPERATING PROCEDURES

By scanning odd numbered lines in one field and even numbered lines in the next

E2B04 How is color information sent in analog SSTV?
Color lines are sent sequentially

E2B05 Vestigial sideband in analog fast-scan TV transmissions
reduces bandwidth while increasing fidelity of low frequency video

E2B06 What is vestigial sideband modulation?
Amplitude modulation in which one complete sideband and a portion of the other are transmitted

E2B07 What types of modulation are used for amateur television DVB-T signals?
QAM and QPSK

E2B08 What technique allows commercial analog TV receivers to be used for fast-scan TV on the 70 cm band?
Transmitting on channels shared with cable TV

E2B09 What kind of receiver can be used to receive and decode SSTV using the Digital Radio Mondial (DRM) protocol?
SSB

E2B10 What aspect of an analog slow-scan television signal encodes the brightness of the picture?
Tone frequency

E2B11 What is the function of the Vertical Interval Signaling (VIS) code sent as part of an SSTV transmission?
To identify the SSTV mode being used

E2B12 What signals SSTV receiving software to begin a new picture line?
Specific tone frequencies

E2C Contest and DX Operating

E2C01 What indicator is required to be used by US-licensed operators when operating a station via remote control and the remote transmitter is located in the US?
No additional indicator is required

E2C02 The format used for exchanging amateur radio log data is **ADIF**

E2C03 From What bands is amateur radio contesting generally excluded?
30 meters

E2C04 Which frequencies can be used for amateur radio mesh networks?
Frequencies shared with various unlicensed wireless data services

E2C05 What is the function of a DX QSL Manager?
Handle the receiving and sending of confirmation cards for a DX station

E2C06 During a VHF/UHF contest, in which band segment would you expect to find the highest level of SSB or CW activity?
In the weak signal segment of the band, with most of the activity near the calling frequency

E2C07 What is the Cabrillo format?
A standard for submission of electronic contest logs

E2C08 What contacts may be confirmed through Logbook of the World (LoTW)?
Special Event contacts between stations in the US
Contacts between US and non-US stations
Contacts for Worked All States credit
All these choices are correct

E2C09 What type of equipment is commonly used to implement an amateur radio mesh network?
A wireless router running custom firmware

E2C10 Why might a DX station transmit and receive on different frequencies?
Because the DX station may be transmitting on a frequency that is prohibited to some responding stations
To separate the calling stations from the DX station
To improve operating efficiency by reducing interference
All these choices are correct

E2C11 How should you generally identify your station when attempting to contact a DX station during a contest or in a pileup?
Send your full call sign once or twice

E2C12 What indicates the delay between a control operator action and the corresponding change in the transmitted signal?
Latency

E2D Operating methods: VHF and UHF

E2D01 Which digital mode is designed for meteor scatter communications?
MSK144

E2D02 The information replacing signal-to-noise ratio when using FT8 or FT4 in a VHF contest is:
Grid square

E2D03 What digital modes is designed for EME communications?
Q65

E2D04 What technology is used for real-time tracking of balloons carrying amateur radio transmitters?
APRS

E2D05 What is the characteristic of the JT65 mode?
Decodes signals with a very low signal-to-noise ratio

E2D06 What is a method of establishing EME contacts?
Time-synchronous transmissions alternating between stations

E2D07 What digital protocol is used by APRS?
AX.25

E2D08 What type of packet frame is used to transmit APRS beacon data?
Unnumbered Information

E2D09 What type of modulation is used by JT65?
Multitone AFSK

E2D10 A packet path WIDE3-1 designates:
Three digipeater hops are requested, with one remaining

E2D11 APRS stations relay data
by packet digipeaters

E2E Operating Methods: Digital

E2E01 What types of modulation is common for data emissions below 30 MHz?
FSK

E2E02 What synchronizes WSJT-X digital mode transmit/receive timing?
Synchronization of computer clocks

E2E03 To what does the "4" in FT4 refer?
Four-tone continuous phase frequency shift keying

REVIEW – OPERATING PROCEDURES

E2E04 Which is a characteristic of FST4 mode?
Four-tone Gaussian frequency shift keying
Variable transmit/receive periods
Seven different tone spacings
All these choices are correct

E2E05 Which of these digital modes does not support keyboard-to-keyboard operation?
WSPR

E2E06 The length of an FT8 transmission cycle is
15 seconds

E2E07 How does Q65 differ from JT65?
Multiple receive cycles are averaged

E2E08 HF digital modes used to transfer binary files?
PACTOR

E2E09 Digital mode using variable-length character coding?
PSK31

E2E10 Which of these digital modes has the narrowest bandwidth?
FT8

E2E11 Difference between direct FSK and audio FSK?
FSK modulates the transmitter VFO

E2E12 How do ALE stations establish contact?
ALE constantly scans a list of frequencies, activating the radio when the designated call sign is received

E2E13 Digital modes with the highest data throughput under clear communication conditions?
PACTOR IV

E3 - RADIO WAVE PROPAGATION

[3 Exam Questions - 3 Groups]

E3A Electromagnetic Waves

E3A01 What is the approximate maximum separation measured along the surface of the Earth between two stations communicating by EME?
12,000 miles, if the moon is "visible" by both stations

E3A02 What characterizes libration fading of an EME signal?
A fluttery irregular fading

E3A03 When scheduling EME contacts, which of these conditions will generally result in the least path loss?
When the moon is at perigee

E3A04 In what direction does an electromagnetic wave travel?
At a right angle to the electric and magnetic fields.

E3A05 How are the component fields of an electromagnetic wave oriented?
They are at right angles

E3A06 To continue a long-distance contact when the MUF for that path decreases due to darkness
Switch to a lower frequency HF band

E3A07 Atmospheric ducts capable of propagating microwave signals often form over what geographic feature?
Large bodies of water

E3A08 When a meteor strikes the Earth's atmosphere, a linear ionized region is formed at what region of the ionosphere?
The E region

REVIEW – RADIO WAVE PROPAGATION

E3A09 Frequency ranges most suited for meteor scatter communications?
28 MHz - 148 MHz

E3A10 What determines the speed of electromagnetic waves through a medium?
The index of refraction

E3A11 What is a typical range for tropospheric duct propagation of microwave signals?
100 miles to 300 miles

E3A12 What is most likely to result in auroral propagation?
Severe geomagnetic storms.

E3A13 Emission mode best for auroral propagation?
CW

E3A14 What are circularly polarized electromagnetic waves?
Waves with rotating electric and magnetic fields

E3B Transequatorial Propagation

E3B01 Transequatorial propagation is most likely to occur
Between two points separated by 2,000 miles to 3,000 miles over a path perpendicular to the geomagnetic equator.

E3B02 What is the approximate maximum range for signals using transequatorial propagation?
5000 miles

E3B03 Best time of day for transequatorial propagation?
Afternoon or early evening

E3B04 What are "extraordinary" and "ordinary" waves?
Independently propagating, elliptically polarized waves created in the atmosphere

Page 142   Extra Class – The Easy Way

E3B05 Which path is most likely to support long-distance propagation on 160 meters?
A path entirely in darkness

E3B06 On which amateur bands is long-path propagation most frequent?
40 and 20 meters

E3B07 What effect does lowering a signal's transmitted elevation angle have on ionospheric HF skip propagation ?
The distance covered by each hop increases

E3B08 How does the maximum range of ground-wave propagation change when the signal frequency is increased?
It decreases

E3B09 At what time of year is sporadic E propagation most likely to occur?
Around the solstices, especially the summer solstice

E3B10 The effect of chordal-hop propagation is
The signal experiences less loss compared to multi-hop propagation which uses Earth as a reflector

E3B11 At what time of day is sporadic-E propagation most likely to occur?
Between sunrise and sunset

E3B12 What is chordal hop propagation?
Successive ionospheric refractions without an intermediate reflection from the ground

E3B13 The polarization supported by ground-wave propagation is
Vertical

E3C Radio Propagation

E3C01 The cause of short-term radio blackouts is
Solar flares

REVIEW – RADIO WAVE PROPAGATION

E3C02 What is indicated by a rising A-index or K-index?
Increasing disturbance of the geomagnetic field

E3C03 What signal paths is most likely to experience high levels of absorption when the A-index or K-index is elevated?
Through the auroral oval

E3C04 What does the value of Bz (B sub Z) represent?
North-south strength of the interplanetary magnetic field

E3C05 What orientation of Bz (B sub z) increases the likelihood that incoming particles from the sun will cause disturbed conditions?
Southward

E3C06 How does the VHF/UHF radio horizon compare to the geometric horizon?
Approximately 15 percent farther

E3C07 What indicates the greatest solar flare intensity?
Class X

E3C08 A space-weather term for an extreme geomagnetic strom is **G5**

E3C09 The data reported by an amateur radio propagation reporting network is
Digital-mode and CW signals

E3C10 What does the 304A solar parameter measure?
UV emissions at 304 angstroms, correlated to the solar flux index

E3C11   VOACAP software models **HF propagation**

E3C12 What might be indicated by a sudden rise in radio background noise across a large portion of the HF spectrum?
A coronal mass ejection or a solar flare has occurred

E4 - AMATEUR PRACTICES
[5 Exam Questions - 5 Groups]

E4A Test Equipment

E4A01 What limits the highest frequency signal that can be accurately displayed on a digital oscilloscope?
Sampling rate of the analog-to-digital converter

E4A02 What parameters does a spectrum analyzer display on the vertical and horizontal axes?
Signal amplitude and frequency

E4A03 What test instruments is used to display spurious signals and/or intermodulation distortion products generated by an SSB transmitter?
Spectrum analyzer

E4A04 How is the compensation of an oscilloscope probe typically adjusted?
A square wave is displayed and the probe is adjusted until the horizontal portions of the displayed wave are as nearly flat as possible

E4A05 The purpose of the prescaler function on a frequency counter?
Reduce the signal frequency to within the counter's operating range

E4A06 What is the effect of aliasing on a digital oscilloscope when displaying a waveform?
A false, jittery low-frequency version of the signal is displayed

E4A07 What is an advantage of using an antenna analyzer compared to an SWR bridge?
Antenna analyzers compute SWR and impedance automatically

E4A08 What is used to measure SWR?
Directional wattmeter
Vector network analyzer
Antenna analyzer
All these choices are correct

E4A09 What is good practice when using an oscilloscope probe?
Minimize the length of the probe's ground connection

E4A10 Which trigger mode is most effective when using an oscilloscope to measure a linear power supply's output ripple?
Line

E4A11 What can be measured with an antenna analyzer?
Velocity factor
Cable length
Resonant frequency of a tuned circuit
 All these choices are correct

E4B Measurements

E4B01 What factor most affects the accuracy of a frequency counter?
Time base accuracy

E4B02 What is the significance of voltmeter sensitivity expressed in ohms per volt?
The full scale reading of the voltmeter multiplied by its ohms per volt rating will indicate the input impedance of the voltmeter

E4B03 Which S parameter is equivalent to forward gain?
S21

E4B04 Which S parameter represents input port return loss or reflection coefficient (equivalent to VSWR)?
S11

E4B05 What three test loads are used to calibrate an RF vector network analyzer?
Short circuit, open circuit, and 50 ohms

E4B06 How much power is being absorbed by the load when a directional power meter connected between a transmitter and a terminating load reads 100 watts forward power and 25 watts reflected power?
75 watts

E4B07 What do the subscripts of S parameters represent?
The port or ports at which measurements are made

E4B08 What can be used to measure the Q of a series-tuned circuit?
The bandwidth of the circuit's frequency response

E4B09 What is measured by a two-port vector network analyzer?
Filter frequency response

E4B10 What methods measures intermodulation distortion in an SSB transmitter?
Modulate the transmitter using two AF signals having non-harmonically related frequencies and observe the RF output with a spectrum analyzer

E4B11 Which is measured with a vector network analyzer?
Input impedance
Output impedance
Reflection coefficient
All these choices are correct

E4C Receiver Performance

E4C01 What is an effect of excessive phase noise in an SDR receiver's master clock oscillator?
It can combine with strong signals on nearby frequencies to generate interference

E4C02 What receiver circuits can be effective in eliminating interference from strong out-of-band signals?
A front-end filter or preselector

E4C03 The term for the suppression in an FM receiver of one signal by another stronger signal on the same frequency?
Capture effect

E4C04 What is the noise figure of a receiver?
The ratio in dB of the noise generated by the receiver to the theoretical minimum noise

E4C05 A receiver noise floor of -174 dBm represents?
The theoretical noise in a 1 Hz bandwidth at the input of a perfect receiver at room temperature

E4C06 Increasing a receiver's bandwidth from 50 Hz to 1,000 Hz increases the receiver's noise floor by
13 dB

E4C07 What does the MDS of a receiver represent?
The minimum discernible signal

E4C08 An SDR receiver is overloaded when input signals exceed what level?
The reference voltage of the analog-to-digital converter

E4C09 What choices is a good reason for selecting a high frequency for the design of the IF in a superheterodyne HF or VHF communications receiver?
Easier for front-end circuitry to eliminate image responses

E4C10 What is an advantage of having a variety of receiver IF bandwidths from which to select?
Receive bandwidth can be set to match the modulation bandwidth, maximizing signal-to-noise ratio and minimizing interference

E4C11 Why can an attenuator be used to reduce receiver overload on the lower frequency HF bands with little or no impact on signal-to-noise ratio?
Atmospheric noise is generally greater than internally generated noise even after attenuation

E4C12 How does a narrow-band roofing filter affect receiver performance?
It improves blocking dynamic range by attenuating strong signals near the receive frequency

E4C13 What is reciprocal mixing?
Local oscillator phase noise mixing with adjacent strong signals to create interference to desired signals

E4C14 The purpose of the receiver IF Shift control is
to reduce interference from stations transmitting on adjacent frequencies

E4D Receiver Performance Characteristics

E4D01 The blocking dynamic range of a receiver is?
The difference in dB between the noise floor and the level of an incoming signal that will cause 1 dB of gain compression

E4D02 What describes problems caused by poor dynamic range in a receiver?
Spurious signals caused by cross-modulation and desensitization from strong adjacent signals

E4D03 What creates intermodulation interference between two repeaters in close proximity?
When the signals mix in the final amplifier of one or both transmitters

E4D04 What may reduce or eliminate intermodulation interference in a repeater caused by a nearby transmitter?

A properly terminated circulator at the output of the repeater's transmitter

E4D05 What transmitter frequencies would cause an intermodulation-product signal in a receiver tuned to 146.70 MHz when a nearby station transmits on 146.52 MHz?
146.34 MHz and 146.61 MHz

E4D06 The term for the reduction in receiver sensitivity caused by a strong signal near the received frequency?
Desensitization

E4D07 What reduces the likelihood of receiver desensitization?
Insert attenuation before the first RF stage

E4D08 What causes intermodulation in an electronic circuit?
Nonlinear circuits or devices

E4D09 What is the purpose of the preselector in a communications receiver?
To increase rejection of signals outside the desired band

E4D10 What does a third-order intercept level of 40 dBm mean with respect to receiver performance?
A pair of 40 dBm input signals will theoretically generate a third-order intermodulation product that has the same output amplitude as either of the input signals

E4D11 Odd-order intermodulation products created within a receiver, are of particular interest compared to other products?
Odd-order products of two signals in the band of interest are also likely to be within the band

E4D12 What is the link margin in a system with a transmit power level of 10 W (+40dBm), a system antenna gain of 10dBi, a cable loss of 3 dB, a path loss of 136 dB, a receiver minimum discernable signal of -103 dBm, and a required signal-to-noise ratio of 6 dB?

+8 dB

E4D13 What is the received signal level with a transmit power of 10 W (+40 dBm), a transmit antenna gain of 6 dBi, a receive antenna gain of 3 dBi, and a path loss of 100 dB?
-51 dBm

E4D14 What power level does a receiver minimum discernible signal of -100 dBm represent?
0.1 picowatts

E4E Noise and Interference

E4E01 What problem can occur when using an automatic notch filter (ANF) to remove interfering carriers while receiving CW signals?
Removal of the CW signal as well as the interfering carrier

E4E02 What types of noise can often be reduced with a digital noise reduction?
Broadband white noise
Ignition noise
Power line noise
All these choices are correct

E4E03 What types of noise are removed by a noise blanker?
Impulse noise.

E4E04 How can conducted noise from an automobile battery charging system be suppressed?
By installing ferrite chokes on the charging system leads

E4E05 Suppress radio frequency interference from a line-driven AC motor with?
A brute-force AC-line filter in series with the motor leads

E4E06 A type of electrical interference caused by computer network equipment?
Unstable modulated or unmodulated signals

E4E07 Why do shielded cables radiate or receive interference?
Common mode currents on the shield and conductors

E4E08 What current flows equally on all conductors of an unshielded multi-conductor cable?
Common-mode current

E4E09 An undesirable effect of using a noise blanker?
Strong signals may be distorted and appear to cause spurious emissions

E4E10 What can create intermittent loud roaring or buzzing AC line interference?
Arcing contacts in a thermostatically controlled device
A defective doorbell or doorbell transformer inside a nearby residence
A malfunctioning illuminated advertising display
All these choices are correct

E4E11 What could cause local AM broadcast band signals to combine to generate spurious signals in the MF or HF bands?
Nearby corroded metal connections mixing and re-radiating the broadcast signals

E4E12   What causes interference received as a series of carriers at regular intervals across a wide frequency range?
Switch-mode power supplies

E4E13 Where to install a station AC surge protector?
On the single point ground panel

E4E14 The purpose of a single point ground panel is to
ensure all lightning protectors activate at the same time.

E5 – ELECTRICAL PRINCIPLES

[4 Exam Questions - 4 Groups]

E5A Resonance and Q

E5A01 What can cause the voltage across reactances in a series RLC circuit to be higher than the voltage applied to the entire circuit?
Resonance

E5A02 What is the resonant frequency of an RLC circuit if R is 22 ohms, L is 50 microhenries and C is 40 picofarads?
3.56 MHz

E5A03 What is the magnitude of the impedance of a series RLC circuit at resonance?
Approximately equal to circuit resistance

E5A04 What is the magnitude of the impedance of a parallel RLC circuit at resonance?
Approximately equal to circuit resistance

E5A05 What is the result of increasing the Q of an impedance-matching circuit?
Matching bandwidth is decreased

E5A06 What is the magnitude of the circulating current within the components of a parallel LC circuit at resonance?
It is at a maximum

E5A07 What is the magnitude of the current at the input of a parallel RLC circuit at resonance?
Minimum

E5A08 What is the phase relationship between the current through and the voltage across a series resonant circuit at resonance?
The voltage and current are in phase

E5A09 How is the Q of an RLC parallel resonant circuit calculated?
Resistance divided by the reactance of either the inductance or capacitance

E5A10 What is the resonant frequency of an RLC circuit if R is 33 ohms, L is 50 microhenries and C is 10 picofarads?
7.12 MHz

E5A11 What is the half-power bandwidth of a resonant circuit that has a resonant frequency of 7.1 MHz and a Q of 150?
47.3 kHz

E5A12 What is the half-power bandwidth of a resonant circuit that has a resonant frequency of 3.7 MHz and a Q of 118?
31.4 kHz

E5A13 What is an effect of increasing Q in a series resonant circuit?
Internal voltages increase

E5B Time Constants and Phase

E5B01 What is the term for the time required for the capacitor in an RC circuit to be charged to 63.2% of the applied voltage or to discharge to 36.8% of its initial voltage?
One time constant.

E5B02 What letter is commonly used to represent susceptance?
Letter B

E5B03 How is impedance in polar form converted to an equivalent admittance?
Take the reciprocal of the magnitude and change the sign of the angle

E5B04 What is the time constant of a circuit having two 220-microfarad capacitors and two 1-megohm resistors, all in parallel?
220 seconds

E5B05 What happens to the magnitude of a pure reactance when it is converted to a susceptance?
It is replaced by the reciprocal

E5B06 What is susceptance?
The imaginary part of admittance

E5B07 What is the phase angle between the voltage across and the current through a series RLC circuit if XC is 500 ohms, R is 1 kilohm, and XL is 250 ohms?
14.0 degrees with the voltage lagging the current

E5B08 What is the phase angle between the voltage across and the current through a series RLC circuit if XC is 300 ohms, R is 100 ohms, and XL is 100 ohms?
63 degrees with the voltage lagging the current

E5B09 What is the relationship between the AC current through a capacitor and the voltage across a capacitor?
Current leads voltage by 90 degrees

E5B10 What is the relationship between the AC current through an inductor and the voltage across an inductor?
Voltage leads current by 90 degrees

E5B11 What is the phase angle between the voltage across and the current through a series RLC circuit if XC is 25 ohms, R is 100 ohms, and XL is 75 ohms?
27 degrees with the voltage leading the current

E5B12 What is admittance?
The inverse of impedance

E5C Coordinate Systems and Phasors

E5C01 What represents capacitive reactance of 100 ohms in rectangular notation?
0–j100

E5C02 How are impedances described in polar coordinates?
By magnitude and phase angle

E5C03 What represents a pure inductive reactance in polar coordinates?
A positive 90 degree phase angle

E5C04 What type of Y-axis scale is most often used for graphs of circuit frequency response?
Logarithmic

E5C05 What is the name of the diagram used to show the phase relationship between impedances at a given frequency?
Phasor diagram

E5C06 What does the impedance 50–j25 represent?
50 ohms resistance in series with 25 ohms capacitive reactance

E5C07 Where is the impedance of a pure resistance plotted on rectangular coordinates?
On the horizontal axis

E5C08 What coordinate system is often used to display the phase angle of a circuit containing resistance, inductive and/or capacitive reactance?
Polar coordinates

E5C09 When using rectangular coordinates to graph the impedance of a circuit, what do the axes represent?
The X axis represents the resistive component and the Y axis represents the reactive component

E5C10 Which point on Figure E5-1 best represents the impedance of a series circuit consisting of a 400-ohm resistor and a 38-picofarad capacitor at 14 MHz?
Point 4

E5C11 Which point in Figure E5-1 best represents the impedance of a series circuit consisting of a 300-ohm resistor and an 18-microhenry inductor at 3.505 MHz?
Point 3

E5C12 Which point on Figure E5-1 best represents the impedance of a series circuit consisting of a 300-ohm resistor and a 19-picofarad capacitor at 21.200 MHz?
Point 1

E5D RF Effects

E5D01 What is the result of conductor skin effect?
Resistance increases as frequency increases, because RF current flows, closer to the surface

E5D02 Why is it important to keep lead lengths short for components used in circuits for VHF and above?
To minimize inductive reactance

E5D03 What is the phase relationship between current and voltage for reactive power?
They are 90 degrees out of phase

E5D04 Why are short connections used at microwave frequencies?
To reduce phase shift along the connection

E5D05 What parasitic characteristic causes electrolytic capacitors to be unsuitable for use at RF?
Inductance

REVIEW – ELECTRICAL PRINCIPLES

E5D06 What parasitic characteristic creates an inductor's self-resonance?
Inter-turn capacitance

E5D07 What combines to create the self-resonance of a component?
The components nominal and parasitic reactance.

E5D08 The primary cause of loss in film capacitors at RF is
Skin effect

E5D09 What happens to reactive power in ideal inductors and capacitors?
Energy is stored in magnetic or electric fields, but power is not dissipated.

E5D10 As a conductor's diameter increases, what is the effect on its electrical length?
It increases

E5D11 How much real power is consumed in a circuit consisting of a 100-ohm resistor in series with a 100-ohm inductive reactance drawing 1 ampere?
100 watts

E5D12 What is reactive power?
Wattless, nonproductive power

E6 – CIRCUIT COMPONENTS

6 Exam Questions - 6 Groups]

E6A Semiconductors

E6A01 Gallium arsenide is used as a semiconductor material?
In microwave circuits

E6A02 What semiconductor material contains excess free electrons?
N-type

E6A03 Why does a PN-junction diode not conduct current when reverse biased?
Holes in P-type material and electrons in the N-type material are separated by the applied voltage, widening the depletion region

E6A04 What is the name given to an impurity atom that adds holes to a semiconductor crystal structure?
Acceptor impurity

E6A05 How does DC input impedance at the gate of a field-effect transistor compare with the DC input impedance of a bipolar transistor?
An FET has higher input impedance

E6A06 What is the beta of a bipolar junction transistor?
The change in collector current with respect to the change in base current

E6A07 What indicates that a silicon NPN junction transistor is biased on?
Base-to-emitter voltage of approximately 0.6 to 0.7 volts

E6A08 What term indicates the frequency at which the grounded-base current gain of a transistor has decreased to 0.7 of the gain obtainable at 1 kHz?
Alpha cutoff frequency

E6A09 What is a depletion-mode field effect transistor (FET)?
An FET that exhibits a current flow between source and drain when no gate voltage is applied

E6A10 In Figure E6-1, (at the back of this book) what is the schematic symbol for an N-channel dual-gate MOSFET?
4

E6A11 In Figure E6-1, what is the schematic symbol for a P-channel junction FET?
1

E6A12 What is the purpose of connecting Zener diodes between a MOSFET gate and its source or drain?
To protect the gate from static damage

E6B Diodes

E6B01 The most useful characteristic of a Zener diode?
A constant voltage drop under conditions of varying current

E6B02 A Schottky diode as compared to an ordinary silicon diode when used as a power supply rectifier?
Lower forward voltage drop

E6B03 What property of an LED's semiconductor material determines its forward voltage drop?
Band gap

E6B04 What type of semiconductor device is designed for use as a voltage-controlled capacitor?
Varactor diode

E6B05 Aa PIN diode is useful as an RF switch because of?
Low junction capacitance

E6B06 What is a common use of a Schottky diode?
As a VHF/UHF mixer or detector

E6B07 A junction diode fails from excessive current because?
Excessive junction temperature

E6B08 What is a Schottky barrier diode?
Metal-semiconductor junction

E6B09 What is a common use for point-contact diodes?
As an RF detector

E6B10 In Figure E6-2, (at the back of this book) what is the schematic symbol for a Shottky diode?
6

E6B11 What is used to control the attenuation of RF signals by a PIN diode?
Forward DC bias current

E6C Digital ICs

E6C01 What is the function of hysteresis in a comparator?
To prevent input noise from causing unstable output signals

E6C02 What happens when the level of a comparator's input signal crosses the threshold?
The comparator changes its output state

E6C03 What is tri-state logic?
Logic devices with 0, 1, and high-impedance output states

E6C04 What is an advantage of BiCMOS logic?
It has the high input impedance of CMOS and the low output impedance of bipolar transistors

REVIEW – CIRCUIT COMPONENTS

E6C05 Which digital logic family has the lowest power consumption?
CMOS

E6C06 Why do CMOS digital integrated circuits have high immunity to noise on the input signal or power supply?
The input switching threshold is about half the power supply voltage

E6C07 What best describes a pull-up or pull-down resistor?
A resistor connected to the positive or negative supply line used to establish a voltage when an input or output is an open circuit

E6C08 In Figure E6-3, what is the schematic symbol for a NAND gate?
2

E6C09 What is used to design the configuration of a field-programmable gate array (FPGA)?
Hardware description language (HDL)

E6C10 In Figure E6-3, what is the schematic symbol for a NOR gate?
4

E6C11 In Figure E6-3, what is the schematic symbol for the NOT operation (inversion)?
5

E6D Inductors and Piezoelectricity

E6D01 Piezoelectricity is:
A characteristic of materials that generate a voltage when stressed and that flex when voltage is applied

E6D02 The equivalent circuit of a quartz crystal is:
Series RLC in parallel with a shunt C representing electrode and stray capacitance

E6D03 An aspect of the piezoelectric effect is:
Mechanical deformation of a material due to the application of a voltage

E6D04 Why are the cores of inductors and transformers are constructed of thin layers?
To reduce power loss from eddy currents in the core.

E6D05 How do ferrite and powdered iron compare for use in an inductor core?
Ferrite toroids generally require fewer turns to produce a given inductance value

E6D06 What core material property determines the inductance of an inductor?
Permeability

E6D07 What is current in the primary winding of a transformer called when there is no load on the secondary?
Magnetizing current

E6D08 Which material has the highest temperature stability of its magnetic characteristics?
Powdered-iron

E6D09 What devices are commonly used as VHF and UHF parasitic suppressors at the input and output terminals of a transistor HF amplifier?
Ferrite beads

E6D10 What is a primary advantage of using a toroidal core instead of a solenoidal core in an inductor?
Toroidal cores confine most of the magnetic field within the core material

E6D11 Which type of core material decreases inductance when inserted into a coil?
Brass

E6D12 What causes inductor saturation?
Operating at excessive magnetic flux

E6E Semiconductor Materials and Packages

E6E01 Why is gallium arsenide (GaAs) useful for semiconductor devices operating at UHF and higher frequencies?
Higher electron mobility

E6E02 What device packages is a through-hole type?
DIP

E6E03 What materials is likely to provide the highest frequency of operation when used in MMICs?
Gallium nitride

E6E04 Which is the most common input and output impedance of circuits that use MMICs?
50 ohms

E6E05 What noise figure values is typical of a low-noise UHF preamplifier?
.05 dB

E6E06 What characteristics of the MMIC make it a popular choice for VHF through microwave circuits?
Controlled gain, low noise figure, and constant input and output impedance over the specified frequency range

E6E07 What type of transmission line is used for connections to MMICs?
Microstrip

E6E08 How is power supplied to the most common type of MMIC?
Through a resistor and/or RF choke connected to the amplifier output lead

E6E09 What component package types have the least parasitic effects at frequencies above the HF range?
Surface mount

E6E10 What advantage does surface-mount technology offer at RF compared to using through-hole components?
Smaller circuit area
Shorter circuit-board traces
Components have less parasitic inductance and capacitance
All these choices are correct

E6E11 What is a characteristic of DIP packaging used for integrated circuits?
A total of two rows of connecting pins placed on opposite sides of the package (Dual In-line Package)

E6E12 Why are DIP through-hole package ICs not typically used at UHF and higher frequencies?
Excessive lead length

E6F Electro-Optical Technology

E6F01 What absorbs the energy from light falling on a photovoltaic cell?
Electrons

E6F02 What happens to the to photoconductive material when light shines on it?
Resistance decreases

E6F03 What is the most common configuration of an optoisolator or optocoupler?
An LED and a phototransistor

E6F04 What is the photovoltaic effect?
The conversion of light to electrical energy

E6F05 An optical shaft encoder is?

REVIEW – CIRCUIT COMPONENTS

A device that detects rotation of a control by interrupting a light source with a patterned wheel

E6F06 Material most commonly used to create photoconductive devices?
A crystalline semiconductor

E6F07 What is a solid-state relay?
A device that uses semiconductors to implement the functions of an electromechanical relay

E6F08 Why are optoisolators often used in conjunction with solid-state circuits when switching 120 VAC?
Optoisolators provide an electrical isolation between a control circuit and the circuit being switched

E6F09 What is the efficiency of a photovoltaic cell?
The relative fraction of light that is converted to current

E6F10 What is the most common type of photovoltaic cell used for electrical power generation?
Silicon

E6F11 What is the approximate open-circuit voltage produced by a fully illuminated silicon photovoltaic cell?
0.5 V

E7 – PRACTICAL CIRCUITS

[8 Exam Questions - 8 Groups]

E7A Digital Circuits

E7A01 Which circuit is bistable?
A flip-flop

E7A02 What is the function of a decade counter?
It produces one output pulse for every 10 input pulses

E7A03 What can divide the frequency of a pulse train by 2?
A flip-flop

E7A04 How many flip-flops are required to divide a signal frequency by 16?
4

E7A05 What is a circuit that continuously alternates between two states without an external clock?
Astable multivibrator

E7A06 What is a characteristic of a monostable multivibrator?
It switches temporarily to an alternate state for a set time

E7A07 What logical operation does a NAND gate perform?
It produces a 0 at its output only if all inputs are 1

E7A08 What logical operation does an OR gate perform?
It produces a 1 at its output if any or all inputs are 1

E7A09 What logical operation is performed by a two-input exclusive NOR gate?
It produces 0 at its output if one and only one of its inputs is 1

E7A10 What is a truth table?
A list of inputs and corresponding outputs for a digital device

E7A11 What does positive logic mean?
High voltage represents a 1, low voltage a 0

E7B Amplifiers

E7B01 For what portion of the signal cycle does each active element in a push-pull Class AB amplifier conduct?
More than 180 degrees but less than 360 degrees

E7B02 What is a Class D amplifier?
An amplifier that uses switching technology to achieve high efficiency

E7B03 The circuit required at the output of an
Rf switching amplifier is
A filter to remove harmonic content

E7B04 What is the operating point of a Class A common emitter amplifier?
Approximately halfway between saturation and cutoff

E7B05 What can be done to prevent unwanted oscillations in an RF power amplifier?
Install parasitic suppressors and/or neutralize the stage

E7B06 What is a characteristic of a grounded-grid amplifier?
Low input impedance.

E7B07 What is a likely result when a Class C amplifier is used to amplify a single-sideband phone signal?
Signal distortion and excessive bandwidth

E7B08 Why are switching amplifiers more efficient than linear amplifiers?
The switching device is at saturation or cutoff most of the time

E7B09 A characteristic of an emitter follower (or common collector) amplifier is
Input and output signals are in phase

E7B10 In Figure E7-1, what is the purpose of R1 and R2?
Voltage divider bias

E7B11 In Figure E7-1, what is the purpose of R3?
Self bias

E7B12 What type of amplifier circuit is in Figure E7-1?
Common emitter

E7C Filters and Matching Networks

E7C01 How are the capacitors and inductors of a low-pass filter Pi-network arranged between the network's input and output?
A capacitor is connected between the input and ground, another capacitor is connected between the output and ground, and an inductor is connected between input and output

E7C02 What is the frequency response of a T-network with series capacitors and a shunt inductor?
High-pass

E7C03 What is the purpose of adding an inductor to a Pi-network to create a Pi-L-network?
Greater harmonic suppression

E7C04 How does an impedance-matching circuit transform a complex impedance to a resistive impedance?
It cancels the reactive part of the impedance and changes the resistive part to a desired value

E7C05 Which filter type is described as having ripple in the passband and a sharp cutoff?
A Chebyshev filter

E7C06 What are the distinguishing features of an elliptical filter?
Extremely sharp cutoff with one or more notches in the stop band

E7C07 What describes a Pi-L network?
A Pi-network with an additional output series inductor

E7C08 What is most frequently used as a band-pass or notch filter in VHF and UHF transceivers?
A helical filter

E7C09 What is a crystal lattice filter?
A filter for low-level signals using quartz crystals

E7C10 What filter is used in a 2-meter band repeater duplexer?
A cavity filter

E7C11 What describes a receiving filter's ability to reject signals in adjacent channels?
Shape factor

E7D Power Supplies

E7D01 How does a linear electronic voltage regulator work?
The conduction of a control element is varied to maintain a constant output voltage

E7D02 How does a switchmode voltage regulator work?
By varying the duty cycle of pulses to a filter

E7D03 What device is typically used as a stable voltage reference in a linear voltage regulator?
A Zener diode

E7D04 What describes a three-terminal voltage regulator?
A series regulator

E7D05 Which linear voltage regulator operates by loading the unregulated voltage source?
A shunt regulator

E7D06 What is the purpose of Q1 in the circuit shown in Figure E7-2?
It controls the current to keep the output voltage constant

E7D07 What is the purpose of C2 in the circuit shown in Figure E7-2?
It bypasses rectifier output ripple around D1

E7D08 What type of circuit is shown in Figure E7-2?
Linear voltage regulator

E7D09 Battery operating time is calculated by:
Capacity in amp-hours divided by average current

E7D10 Why is a switching type power supply less expensive and lighter than an equivalent linear power supply?
The high frequency inverter design uses much smaller transformers and filter components for an equivalent power output

E7D11 What is the purpose of an inverter connected to a solar panel output?
Convert the panel's output from DC to AC

E7D12 What is the drop-out voltage of an linear voltage regulator?
Minimum input-to-output voltage required to maintain regulation

E7D13 What is the equation for calculating power dissipated by a series linear voltage regulator?
Voltage difference from input to output multiplied by output current

E7D14 What is the purpose of connecting equal-value resistors across power supply filter capacitors connected in series?
Equalize the voltage across each capacitor
Discharge the capacitors when voltage is removed
Provide a minimum load on the supply
All these choices are correct

E7D15 What is the purpose of a step-start circuit in a high-voltage power supply?
To allow the filter capacitors to charge gradually

E7E Modulation and Demodulation

E7E01 What can be used to generate FM phone emissions?
Reactance modulation of a local oscillator

E7E02 What is the function of a reactance modulator?
To produce PM or FM signals by varying a capacitance

E7E03 What is a frequency discriminator?
A circuit for detecting FM signals

E7E04 A single-sideband phone signal can be generated;
By a balanced modulator followed by a filter

E7E05 What is added to an FM speech channel to boost the higher audio frequencies?
A pre-emphasis network

E7E06 Why is de-emphasis commonly used in FM communications receivers?
For compatibility with transmitters using phase modulation

E7E07 What is meant by the term "baseband" in radio communications?
The frequency range occupied by a message signal prior to modulation

E7E08 What are the principal frequencies that appear at the output of a mixer circuit?
The two input frequencies along with their sum and difference frequencies

E7E09 What occurs when the input signal levels to a mixer are too high?
Spurious mixer products are generated

E7E10 How does a diode envelope detector function?
By rectification and filtering of RF signals

E7E11 Which type of detector circuit is used for demodulating SSB signals?
Product detector.

E7F Software Defined Radio and DSP

E7F01 "Direct sampling" in software defined radios means?
Incoming RF is digitized by an analog-to-digital converter without being mixed with a local oscillator signal

E7F02 What kind of digital signal processing audio filter is used to remove unwanted noise from a received SSB signal?
An adaptive filter

E7F03 What type of digital signal processing filter is used to generate an SSB signal?
A Hilbert-transform filter

E7F04 What method generates an SSB signal using digital signal processing?
Signals are combined in quadrature phase relationship

E7F05 How frequently must an analog signal be sampled to be accurately reproduced?
At least twice the rate of the highest frequency component of the signal

E7F06 What is the minimum number of bits required to sample a signal with a range of 1 volt at a resolution of 1 millivolt?
10 bits

E7F07 What function is performed by a Fast Fourier Transform?
Converting digital signals from the time domain to the frequency domain

E7F08 What is the function of decimation?
Reducing the effective sample rate by removing samples

E7F09 Why is an anti-aliasing digital filter required in a digital decimator?
It removes high-frequency signal components that would otherwise be reproduced as lower frequency components

E7F10 What aspect of receiver analog-to-digital conversion determines the maximum receive bandwidth of a Direct Digital Conversion SDR?
Sample rate

E7F11 What sets the minimum detectable signal level for a direct-sampling SDR receiver in the absence of atmospheric or thermal noise?
Reference voltage level and sample width in bits

E7F12 What is true of Finite Impulse Response (FIR) filters?
FIR filters can delay all frequency components of the signal by the same amount

E7F13 What is the function of taps in a digital signal processing filter?
Provide incremental signal delays for filter algorithms

E7F14 What would allow a digital signal processing filter to create a sharper filter response?
More taps

E7G Operational Amplifiers

E7G01 What is the typical output impedance of an op-amp?
Very low

E7G02 If a capacitor is added across the feedback resistor in Figure E7-3, the frequency response is
A low-pass filter

E7G03 What is the typical input impedance of an op-amp?
Very high

E7G04 What is meant by the term "op-amp input offset voltage"?
The differential input voltage needed to bring the open loop output voltage to zero

E7G05 How can unwanted ringing and audio instability be prevented in an op-amp RC audio filter circuit?
Restrict both gain and Q

E7G06 What is the gain-bandwidth of an operational amplifier?
The frequency at which the open-loop gain of the amplifier equals one

E7G07 What magnitude of voltage gain can be expected from the circuit in Figure E7-3 when R1 is 10 ohms and RF is 470 ohms?
47

E7G08 How does the gain of an ideal operational amplifier vary with frequency?
It does not vary with frequency

E7G09 What will be the output voltage of the circuit shown in Figure E7-3 if R1 is 1000 ohms, RF is 10,000 ohms, and 0.23 volts DC is applied to the input?
-2.3 volts

E7G10 What absolute voltage gain can be expected from the circuit in Figure E7-3 when R1 is 1800 ohms and RF is 68 kilohms?
38

E7G11 What absolute voltage gain can be expected from the circuit in Figure E7-3 when R1 is 3300 ohms and RF is 47 kilohms?
14

E7G12 What is an operational amplifier?
A high-gain, direct-coupled differential amplifier with very high input impedance and very low output impedance

E7H Oscillators and Signal Sources

E7H01 What are three common oscillator circuits?
Colpitts, Hartley and Pierce

E7H02 What is a microphonic?
Changes in oscillator frequency caused by mechanical vibration

E7H03 A phase-lock loop is
An electronic servo loop consisting of a phase detector, a low-pass filter, a voltage-controlled oscillator, and a stable reference oscillator.

E7H04 How is positive feedback supplied in a Colpitts oscillator?
Through a capacitive divider

E7H05 How is positive feedback supplied in a Pierce oscillator?
Through a quartz crystal

E7H06 Functions performed by a phase-locked loop include:
Frequency synthesis and FM demodulation

E7H07 How can an oscillator's microphonic responses be reduced?
Mechanically isolate the oscillator circuitry from its enclosure

E7H08 What components can be used to reduce thermal drift in crystal oscillators?
NP0 capacitors

E7H09 What type of frequency synthesizer circuit uses a phase accumulator, lookup table, digital to analog converter, and a low-pass anti-alias filter?
A direct digital synthesizer

E7H10 What information is contained in the lookup table of a direct digital synthesizer (DDS)?
Amplitude values that represent the desired waveform

E7H11 What are the major spectral impurity components of direct digital synthesizers?
Spurious signals at discrete frequencies

REVIEW – PRACTICAL CIRCUITS

E7H12 What ensures that a crystal oscillator operates on the frequency specified by the crystal manufacturer?
Provide the crystal with a specified parallel capacitance

E7H13 What is a technique for providing highly accurate and stable oscillators needed for microwave transmission and reception?
Use a GPS signal reference
Use a rubidium stabilized reference oscillator
Use a temperature-controlled high Q dielectric resonator
All these choices are correct

E8 - SIGNALS AND EMISSIONS

4 Exam Questions - 4 Groups]

E8A Fourier Analysis, RMS, RF Power

E8A01 What technique shows that a square wave is made up of a sine wave plus all its odd harmonics?
Fourier analysis

E8A02 What is a type of analog-to-digital conversion?
Successive approximation

E8A03 What describes a signal in the time domain?
Amplitude at different times

E8A04 What is "dither" with respect to analog-to-digital converters?
A small amount of noise added to the input signal to reduce quantization noise

E8A05 The benefit of making voltage measurements with a true-RMS calculating meter is
RMS is measured for both sinusoidal and non-sinusoidal signals.

E8A06 What is the approximate ratio of PEP-to-average power in a typical single-sideband phone signal?
2.5 to 1

E8A07 What determines the PEP-to-average power ratio of a single-sideband phone signal?
Speech characteristics

E8A08 Why are direct or flash conversion analog-to-digital converter used for a software defined radio?
Very high speed allows digitizing high frequencies

E8A09 How many different input levels can be encoded by an analog-to-digital converter with 8-bit resolution?
256

E8A10 What is the purpose of a low-pass filter used in conjunction with a digital-to-analog converter?
Remove spurious sampling artifacts from the output signal

E8A11 What is a measure of the quality of an analog-to-digital converter
Total harmonic distortion

E8B Modulation and Demodulation

E8B01 What is the modulation index of an FM signal?
The ratio of frequency deviation to modulating signal frequency

E8B02 How does the modulation index of a phase-modulated emission vary with RF carrier frequency?
It does not depend on the RF carrier frequency

E8B03 What is the modulation index of an FM-phone signal having a maximum frequency deviation of 3000 Hz either side of the carrier frequency when the modulating frequency is 1000 Hz?
3

E8B04 What is the modulation index of an FM-phone signal having a maximum carrier deviation of plus or minus 6 kHz when modulated with a 2 kHz modulating frequency?
3

E8B05 What is the deviation ratio of an FM-phone signal having a maximum frequency swing of plus-or-minus 5 kHz when the maximum modulation frequency is 3 kHz?
1.67

E8B06 What is the deviation ratio of an FM-phone signal having a maximum frequency swing of plus or minus 7.5 kHz when the maximum modulation frequency is 3.5 kHz?
2.14

E8B07 Orthogonal Frequency Division Multiplexing (OFDM) is a technique used for which type of amateur communication?
Digital modes

E8B08 What describes Orthogonal Frequency Division Multiplexing?
A digital modulation technique using subcarriers at frequencies chosen to avoid intersymbol interference

E8B09 What is deviation ratio?
The ratio of the maximum carrier frequency deviation to the highest audio modulating frequency

E8B10 What is frequency division multiplexing (FDM?
Dividing the transmitted signal into separate frequency bands that each carry a different data stream.

E8B11 What is digital time division multiplexing?
Two or more signals are arranged to share discrete time slots of a data transmission

E8C Digital Signals

E8C01 Quadrature Amplitude Modulation (QAM) is
Transmitting data by modulating the amplitude of two carriers of the same frequency but 90 degrees out of phase.

E8C02 What is the definition of symbol rate in a digital transmission?
The rate at which the waveform changes to convey information

E8C03 Why should the phase of a PSK signal be changed at the zero crossing of the RF signal?
To minimize bandwidth

E8C04 What technique minimizes the bandwidth of a PSK31 signal?
Use of sinusoidal data pulses

E8C05 What is the approximate bandwidth of a 13-WPM International Morse Code transmission?
52 Hz

E8C06 What is the bandwidth of an FT8 signal?
50 Hz

E8C07 What is the bandwidth of a 4,800-Hz frequency shift, 9,600 baud ASCI FM transmission?
15.36 kHz

E8C08 How does ARQ accomplish error correction?
If errors are detected, a retransmission is requested

E8C09 Which digital code allows only one bit to change between sequential code values?
Gray code

E8C10 How may data rate be increased without increasing bandwidth?
Using a more efficient digital code

E8C11 What is the relationship between symbol rate and baud?
They are the same

E8C12 What factors affect the bandwidth of a transmitted CW signal?
Keying speed and shape factor (rise and fall time)

E8C13 What is described by the constellation diagram of a QAM or QPSK signal?
The possible phase and amplitude states for each symbol

E8C14 What type of addresses do nodes have in a mesh network?
Internet protocol (IP)

E8C15 What techniques do individual nodes use to form a mesh network?
Discovery and link establishment protocols

E8D Keying defects and Overmodulation

E8D01 Why are received spread spectrum signals resistant to interference?
Signals not using the spread spectrum algorithm are suppressed in the receiver

E8D02 What spread spectrum communications technique uses a high-speed binary bit stream to shift the phase of an RF carrier?
Direct sequence

E8D03 Describe spread spectrum frequency hopping.
Rapidly varying the frequency of a transmitted signal according to a pseudorandom sequence

E8D04 What is the primary effect of extremely short rise or fall time on a CW signal?
The generation of key clicks

E8D05 What is the most common method of reducing key clicks?
Increase keying waveform rise and fall times

E8D06 What is the advantage of including parity bits in ASCII characters?
Some types of errors can be detected

E8D07 What is a common cause of overmodulation of AFSK signals?
Excessive transmit audio levels

E8D08 What parameter evaluates distortion of an AFSK signal caused by excessive input audio levels?
Intermodulation Distortion (IMD)

E8D09 What is considered an acceptable maximum IMD level for an idling PSK signal?
-30 dB

E8D10 What are some of the differences between the Baudot digital code and ASCII?
Baudot uses 5 data bits per character, ASCII uses 7 or 8; Baudot uses 2 characters as letters/figures shift codes, ASCII has no letters/figures shift code

E8D11 What is one advantage of using ASCII code for data communications?
It is possible to transmit both upper and lower case text

E9 - ANTENNAS AND TRANSMISSION LINES

[8 Exam Questions - 8 Groups]

E9A Basic Antenna Parameters

E9A01 What is an isotropic radiator?
A hypothetical, lossless antenna having equal radiation intensity in all directions used as a reference antenna.

E9A02 What is the effective radiated power relative to a dipole of a repeater station with 150 watts transmitter power output, 2 dB feed line loss, 2.2 dB duplexer loss, and 7 dBd antenna gain?
286 watts

E9A03 The term describing total radiated power that takes into account all gains and losses
Effective radiated power

E9A04 Feed point impedance of an antenna is affected by
Antenna height

E9A05 "Ground gain" means
An increase in signal strength from ground reflections

E9A06 The effective radiated power relative to a dipole of a repeater station with 200 watts transmitter power output, 4 dB feed line loss, 3.2 dB duplexer loss, 0.8 dB circulator loss, and 10 dBd antenna gain is?
317 watts

E9A07 The effective isotropic radiated power of a repeater station with 200 watts transmitter power output, 2 dB feed line loss, 2.8 dB duplexer loss, 1.2 dB circulator loss, and 7 dBi antenna gain is?
252 watts

E9A08 The frequency band with the smallest first Fresnel zone is
5.8 Ghz

E9A09 What is antenna efficiency?
Radiation resistance divided by total resistance

E9A10 What improves the efficiency of a ground-mounted quarter-wave vertical antenna?
Installing a ground radial system

E9A11 What determines ground losses for a ground-mounted vertical antenna operating on HF?
Soil conductivity

E9A12 How much gain does an antenna have compared to a half-wavelength dipole if it has 6 dB gain over an isotropic radiator?
3.85 dB

E9B Antenna Patterns

E9B01 What is the 3 dB beamwidth of the antenna radiation pattern shown in Figure E9-1,?
50 degrees

E9B02 What is the front-to-back ratio of the antenna radiation pattern shown in Figure E9-1 ?
18 dB

E9B03 What is the front-to-side ratio of the antenna radiation pattern shown in Figure E9-1, ?
14 dB

E9B04 What is the front-to-back ratio of the radiation pattern shown in Figure E9-2?
28 dB

E9B05 What type of antenna pattern is shown in Figure E9-2?
Elevation

E9B06 What is the elevation angle of peak response in the antenna radiation pattern shown in Figure E9-2?
7.5 degrees

E9B07 What is the difference in radiated power between a lossless antenna with gain and an isotropic radiator driven by the same power?
They are the same

E9B08 What is the far field of an antenna?
The region where the shape of the antenna pattern is independent of distance

E9B09 What type of computer program technique is commonly used for modeling antennas?
Method of Moments

E9B10 What is the principle of a Method of Moments analysis?
A wire is modeled as a series of segments, each having a uniform value of current

E9B11 The disadvantage of decreasing the number of wire segments in an antenna model below 10 segments per half-wavelength is?
The computed feed point impedance may be incorrect

E9C Practical Wire Antennas

E9C01 What is the radiation pattern of two 1/4-wavelength vertical antennas spaced 1/2-wavelength apart and fed 180 degrees out of phase?
A figure-8 oriented along the axis of the array

E9C02 What is the radiation pattern of two 1/4-wavelength vertical antennas spaced 1/4 wavelength apart and fed 90 degrees out of phase?
Cardioid

E9C03 What is the radiation pattern of two 1/4-wavelength vertical antennas spaced 1/2-wavelength apart and fed in phase?
A figure-8 broadside to the axis of the array

E9C04 What happens to the radiation pattern of an unterminated long wire antenna as the wire length is increased?
Additional lobes form with major lobes increasingly aligned with the axis of the antenna

E9C05 What is the purpose of feeding an off-center-fed dipole (OCFD) between the center and one end instead of at the midpoint?
To create a similar feed point impedance on multiple bands

E9C06 What is the effect of adding a terminating resistor to a rhombic or long wire antenna?
It changes the radiation pattern from bidirectional to unidirectional

E9C07 What is the approximate feed point impedance at the center of a two-wire folded dipole antenna?
300 ohms

E9C08 What is a folded dipole antenna?
A half-wave dipole with an additional parallel wire connecting its two ends

E9C09 What describes a G5RV antenna?
A multi-band dipole antenna fed with coax and a balun through a selected length of open wire transmission line

E9C10 What describes a Zepp antenna?
An end-fed half-wavelength dipole antenna

E9C11 How is the far-field elevation pattern of a vertically polarized antenna affected by being mounted over seawater versus soil?
Radiation at low angles increases

E9C12 What describes an extended double Zepp antenna?
A center-fed 1.25-wavelength antenna

E9C13 How does the radiation pattern of a horizontally polarized antenna vary with increasing height above ground?
The takeoff angle of the lowest elevation lobe decreases

E9C14 How does the radiation pattern of a horizontally polarized antenna mounted above a long slope compare with the same antenna mounted above flat ground?
The main lobe takeoff angle decreases in the downhill direction

E9D Directional and Short Antennas

E9D01 How much does the gain of an ideal parabolic dish antenna increase when the operating frequency is doubled?
6 dB

E9D02 How can two linearly polarized Yagi antennas be used to produce circular polarization?
Arrange two Yagis perpendicular to each other with the driven elements at the same point on the boom fed 90 degrees out of phase

E9D03 The most efficient location for a loading coil on an electrically short whip would be
Near the center of the vertical radiator

E9D04 Why should an antenna loading coils have a high ratio of reactance to resistance?
To maximize efficiency

E9D05 Approximately how long is a Yagi's driven element?
1/2 wavelength

E9D06 What happens to the SWR bandwidth when one or more loading coils are used to resonate an electrically short antenna?
It is decreased

E9D07 What is an advantage of using top loading in a shortened HF vertical antenna?
Improved radiation efficiency

E9D08 What happens as the Q of an antenna increases?
SWR bandwidth decreases

E9D09 What is the function of a loading coil used as part of an HF mobile antenna?
To resonate the antenna by canceling capacitive reactance

E9D10 How does the radiation resistance of a base-fed whip antenna change below its resonant frequency?
The radiation resistance decreases

E9D11 Why do most two-element Yagis with normal spacing have a reflector instead of a director?
Higher gain

E9D12 What is the purpose of making a Yagi's parasitic elements either longer or shorter than resonance?
Control of phase shift

E9E Impedance Matching

E9E01 What matching system for Yagi antennas requires the driven element to be insulated from the boom?
Beta or hair pin

E9E02 What antenna matching system matches coaxial cable to an antenna by connecting the shield to the center of the

antenna and the conductor a fraction of a wavelength to one side
The gamma match

E9E03 What matching system uses a section of transmission line connected in parallel with the feed line at or near the feed point?
The stub match

E9E04 What is the purpose of the series capacitor in a gamma-type antenna matching network?
To cancel unwanted inductive reactance

E9E05 What driven element feed point impedance is required to use a beta or hairpin matching system?
Capacitive (driven element shorter than 1/2 wavelength

E9E06 Which of these feed line impedances would be suitable for constructing a quarter-wave Q-section for matching a 100-ohm loop to 50-ohm feed line?
75 ohms

E9E07 What parameter describes the interactions at the load and transmission line?
Reflection coefficient

E9E08 What is a use for a Wilkinson divider?
It is used to divide power equally between two 50-ohm loads while maintaining 50-ohm input impedance

E9E09 What is used to shunt-feed a grounded tower at its base?
Gamma match

E9E10 Which matching system places an inductance across the feed point of a vertical monopole antenna?
Beta or hairpin Question deleted

E9E11 What is the purpose of using multiple driven elements connected through a phasing line?
To control the antenna's radiation pattern

E9F Transmission Lines

E9F01 What is the velocity factor of a transmission line?
The velocity of the wave in the transmission line divided by the velocity of light in a vacuum

E9F02 What has the biggest effect on the velocity factor of a transmission line?
The insulating dielectric material

E9F03 Why is the electrical length of a coaxial cable transmission line longer than its electrical length?
Electrical waves move more slowly in a coaxial cable than in air

E9F04 What impedance does a 1/2-wavelength transmission line present to an RF generator when the line is shorted at the far end?
Very low impedance

E9F05 What is microstrip?
Precision printed circuit conductors above a ground plane that provide constant impedance interconnects at microwave frequencies.

E9F06 What is the approximate physical length of an air-insulated, parallel conductor transmission line that is electrically 1/2 wavelength long at 14.10 MHz?
10.6 meters

E9F07 How does parallel conductor transmission line compare to coaxial cable with plastic dielectric?
Lower loss

E9F08 What is a significant difference between foam dielectric coaxial cable and solid dielectric cable, assuming all other parameters are the same?
Foam dielectric has lower safe operating voltage limits
Lower loss per unit of length
Higher velocity factor
All these choices are correct

E9F09 What impedance does a 1/4-wavelength transmission line present to a generator when the line is shorted at the far end?
Very high impedance

E9F10 What impedance does a 1/8-wavelength transmission line present to a generator when the line is shorted at the far end?
An inductive reactance

E9F11 What impedance does a 1/8-wavelength transmission line present to a generator when the line is open at the far end?
A capacitive reactance

E9F12 What impedance does a 1/4-wavelength transmission line present to a generator when the line is open at the far end?
Very low impedance

E9G Smith Chart

E9G01 What can be calculated using a Smith chart?
Impedance along transmission lines

E9G02 What type of coordinate system is used in a Smith chart?
Resistance circles and reactance arcs

E9G03 What is often determined using a Smith chart?
Impedance and SWR values in transmission lines

REVIEW – ANTENNAS AND TRANSMISSION LINES

E9G04 What are the two families of circles and arcs that make up a Smith chart?
Resistance and reactance

E9G05 What is a common use for a Smith chart?
Determine the length and position of an impedance matching stub

E9G06 On the Smith chart shown in Figure E9-3, what is the name for the large outer circle on which the reactance arcs terminate?
Reactance axis

E9G07 On the Smith chart shown in Figure E9-3, what is the only straight line shown?
The resistance axis

E9G08 How is a Smith chart normalized?
Reassign the prime center's impedance value

E9G09 The third family of circles added to a Smith chart during the process of designing impedance matching networks?
Constant-SWR circles

E9G10 What do the arcs on a Smith chart represent?
Points with constant reactance

E9G11 Wavelength scales on a Smith chart are calibrated?
In fractions of transmission line electrical wavelength

E9H Receiving Antennas

E9H01 When constructing a Beverage antenna, to achieve good performance at the desired frequency?
It should be at least one wavelength long

E9H02 With 160 meter and 80 meter receiving antennas?
Atmospheric noise is so high that directivity is more important than losses.

E9H03 What is receiving directivity factor (RDF)?
Peak antenna gain compared to average gain over the hemisphere around and above the antenna.

E9H04 What is an purpose of placing an electrostatic shield around a small loop direction-finding antenna?
It eliminates unbalanced capacitive coupling to the surroundings, improving the depth of its nulls.

E9H05 What is the challenge of a small wire-loop antenna for direction finding?
It has a bidirectional null pattern

E9H06 What is the correct value of terminating resistance for a Beverage antenna?
Minimum variation in SWR over the desired frequency range.

E9B07 A Beverage antenna's terminating resistor is
to absorb signals from the reverse direction.

E9H08 What is the function of a sense antenna?
It modifies the pattern of a DF antenna array to provide a null in one direction

E9H09 What type of radiation pattern is created by a single-turn terminated loop such as a pennant antenna?
Cardoid

E9H10 How can the output voltage of a multiple-turn receiving loop antenna be increased?
By increasing the number of turns and/or the area enclosed by the loop

E9H11 What feature of a cardioid pattern antenna makes it useful for direction finding?
A single null

E0 – SAFETY

[1 exam question – 1 group]

E0A Safety

E0A01 What is the primary function of an external earth connection or ground rod?
Lightning charge dissipation

E0A02 When evaluating RF exposure levels from your station at a neighbor's home, what must you do?
Ensure signals from your station are less than the uncontrolled Maximum Permitted Exposure (MPE) limits

E0A03 Over what range of frequencies are the FCC human body RF exposure limits most restrictive?
30 to 300 MHz

E0A04 When evaluating a site with multiple transmitters operating at the same time, the operators and licensees of which transmitters are responsible for mitigating over-exposure situations?
Each transmitter that produces 5 percent or more of its MPE limit in areas where the total MPE limit is exceeded.

E0A05 What hazard is created at microwave frequencies?
The high gain antennas commonly used can result in high exposure levels

E0A06 Why are there separate electric (E) and magnetic (H) MPE limits at frequencies below 300 MHz?
The body reacts to electromagnetic radiation from both the E and H fields
Ground reflections and scattering cause the field strength vary with location,
E field and H field radiation intensity peaks can occur at different locations
All these choices are correct

E0A07 "100% tie-off" regarding tower safety means?
At least one lanyard attached to the tower at all times.

E0A08 What does SAR measure?
The rate at which RF energy is absorbed by the body

E0A09 What equipment is exempt from the RF exposure evaluations?
Hand-held transceivers sold before May 3, 2021

E0A10 When must an RF exposure evaluation be performed on an amateur station operating on 80 meters?
An evaluation must always be performed.

E0A11 To what should lanyards be attached when climbing?
Tower legs

E0A12 Where should a shock-absorbing lanyard be attached to a tower when working above ground?
Above the climber's head level

~~~~End of question pool text~~~~

NOTE: Graphics required for questions in sections E5, E6, E7, and E9 are on the following pages.

Figure E5-1

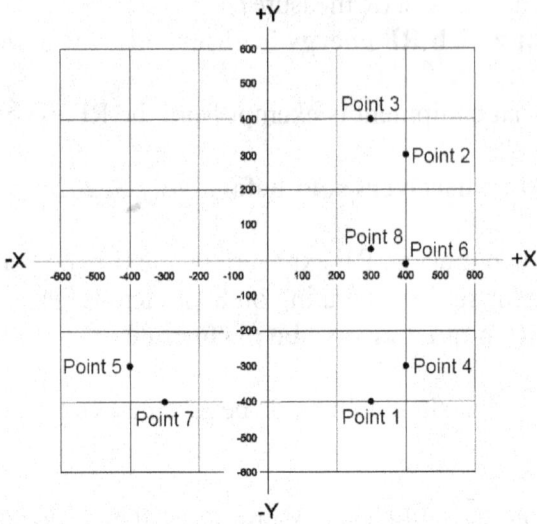

Figure E6-1

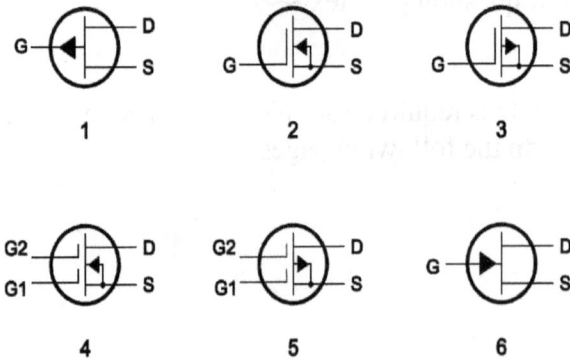

Figure E6-2

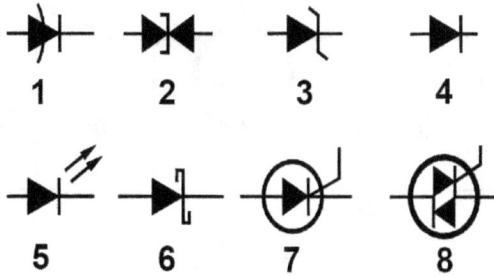

```
  1        2        3        4

  5        6        7        8
```

Figure E6-3

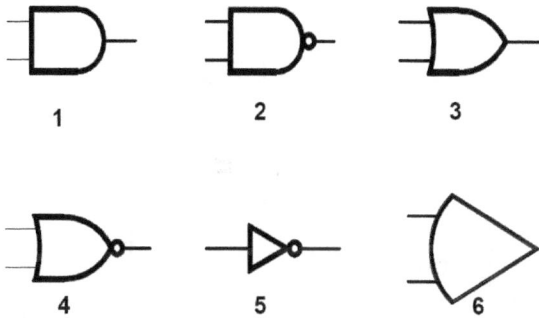

```
  1        2        3

  4        5        6
```

Figure E7-1

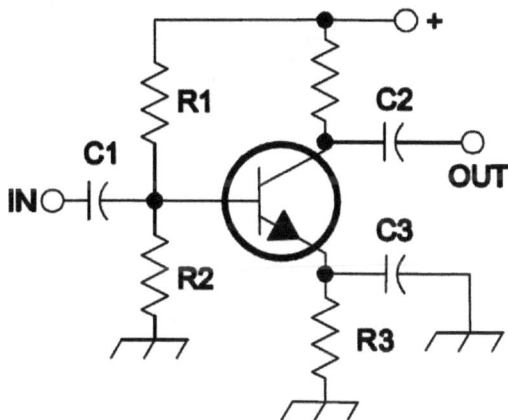

Figure E7-2

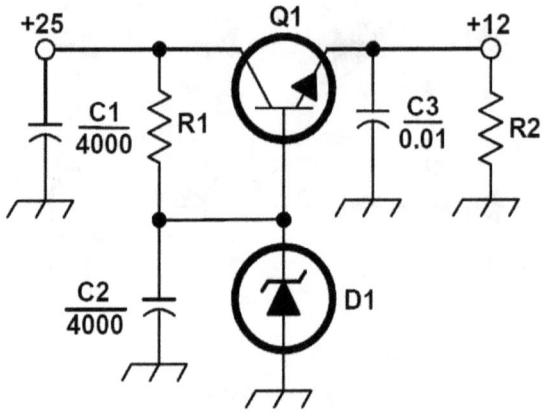

Figure E7-3

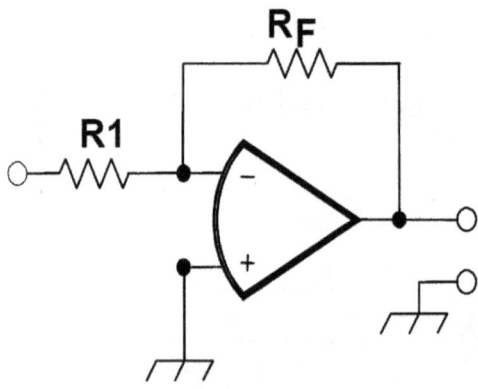

Figure E9-1

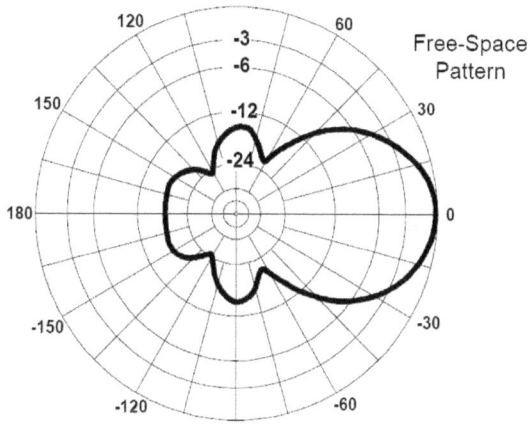

Free-Space Pattern

Figure E9-2

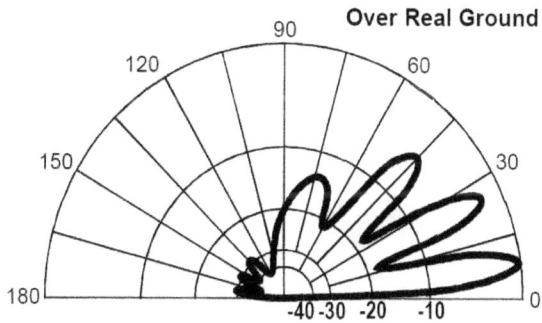

Over Real Ground

Figure E9-3

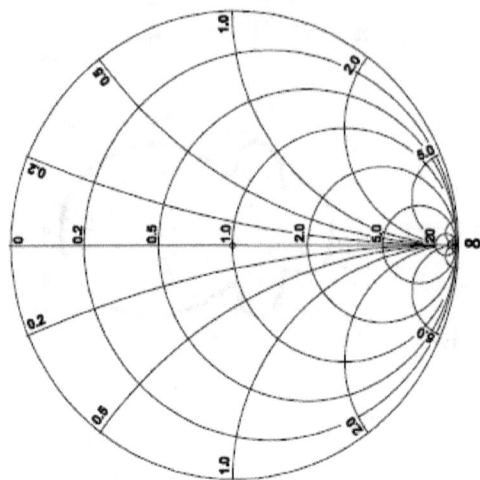

BONUS MATERIAL LEARNING CW

CW is not required for any license class, but is worth learning for its many advantages over voice modes. The advantages include the ability to copy at very low signal levels and use narrower filters. The retro-thrill of the original radio mode is a bonus.

Learning it is not difficult. It is a foreign language with only 26 words and counting to 10. I learned Morse code, and I can't memorize anything. Thankfully, learning Morse is not memorizing. You are training to recognize a sound.

Boy Scouts was my first CW exposure. I had to relearn it as a ham, because in Boy Scouts, CW was audible or a flashing light interpreted to dots and dashes. The decoding was visual, using a chart to look up every letter. That is a very slow multi-step process. It is like translating from English to Spanish to get to French.

Radio CW is audible. Learn to recognize the sound, not memorize dots and dashes. "A" is not short-long or dot-dash. It is not even dit-dah. It is the sound of dit-dah.

Do not learn that "A" sounds like "Ah-pull" or that the letter "A" has a short line and a long line. I've seen pictograms like a child's alphabet block with a picture of a bee for "B." These gimmicks introduce additional mental steps. Working with the sound directly eliminates all the in-between translations.

Learn CW by hearing it in one or two letters at a time until you can immediately recognize the sound as the letter. Then, add more letters and build. This is called the Koch method.

LEARNING CW

The way we sent CW was another impediment to learning. At slow speeds, "A" sounded like diiiiiit-daaaaaaaaaah. Then, the next letter came immediately, with no time to decode. The modern way is called Farnsworth timing.

Farnsworth timing sends the letters at 16-20 words per minute,[4] so each letter has one distinct sound. Don't listen for individual dits and dahs. The letter is one sound.

To slow down the pace, Farnsworth increases the space between letters and words. Increasing the space does two things. It reinforces a single sound as the letter, and it gives the brain extra time between letters to translate. Learn to send using Farnsworth timing. That is how the other guy learned to receive. I send around 22-26 words per minute and slow down by increasing spaces.

After identifying the letters, word recognition comes automatically. You see words, not letters, when reading. "Name" is not "N-A-M-E." The same happens with Morse code. Your call sign, RST, 5NN, TU, 73, and other common "words" come without thinking about the individual letters or the elements that make up the letters.

Learning CW takes practice. In the beginning, listen to a code practice CD or audio file. K7QO offers a free course download on his website, K7QO.net. G4FON has a Koch trainer at G4FON.net.

For more challenging practice, tune in to the W1AW Code Practice Sessions. These are texts from *QST*

---

[4] Words per minute (WPM) is based on a 5 letter word. "Paris" is an often-used standard. Send "Paris" 20 times in a minute for 20 WPM.

magazine, and because the words are longer and not always as predictable, they are harder to copy. ARRL offers code proficiency certificates that will look beautiful on your wall. http://www.arrl.org/w1aw-operating-schedule.

Once you get the letters down (mostly), listen to QSOs. Concentrate on the QSO Trinity: RST, QTH, name. There is a pattern, so you know what to expect. Then, GET ON THE AIR! The best way to practice is with actual contacts. Real QSOs are exciting and won't seem tedious. With time, venture into more complex conversations. Try to match the speed of the other guy, but if you can't, "QRS" means "slow down," and "QRQ" means "speed up."

FISTS CW Club and Straight Key Century Club (SKCC) promote CW and assign you a member number to exchange with other members and collect awards. It makes a fun challenge.

Suggested frequencies to find slower CW include:
3.550 – 3.570 MHz   21.055 – 21.060 MHz
7.055 – 7.060 MHz   28.055 – 28.060 MHz
14.055 – 14.060 MHz
You can also venture into the old Novice Bands and find slow speed CW.

Give CW a try. The world awaits.

INDEX

INDEX

www.ingramcontent.com/pod-product-compliance
Lightning Source LLC
Chambersburg PA
CBHW062130020426
42335CB00013B/1170